AF389159

# Modeling and Simulation of Electricity Systems for Transport and Energy Storage

# Modeling and Simulation of Electricity Systems for Transport and Energy Storage

Editors

**Regina Lamedica**
**Alessandro Ruvio**

MDPI • Basel • Beijing • Wuhan • Barcelona • Belgrade • Manchester • Tokyo • Cluj • Tianjin

*Editors*

Regina Lamedica
Department of Astronautical,
Elecritical and Energetics
Engineering Via Eudossiana
Sapienza—University of Rome
Italy

Alessandro Ruvio
Department of Astronautical,
Elecritical and Energetics
Engineering Via Eudossiana
Sapienza—University of Rome
Italy


*Editorial Office*
MDPI
St. Alban-Anlage 66
4052 Basel, Switzerland


This is a reprint of articles from the Special Issue published online in the open access journal *Energies* (ISSN 1996-1073) (available at: https://www.mdpi.com/journal/energies/special_issues/Modeling_Simulation_Electricity_Systems_Transport_Energy_Storage).

For citation purposes, cite each article independently as indicated on the article page online and as indicated below:

LastName, A.A.; LastName, B.B.; LastName, C.C. Article Title. *Journal Name* **Year**, *Volume Number*, Page Range.

**ISBN 978-3-0365-0326-4 (Hbk)**
**ISBN 978-3-0365-0327-1 (PDF)**

# Contents

# About the Editors

**Regina Lamedica** received her B. Eng. (Hons.) degree in Electrical Engineering from the University of Bologna, Bologna, Italy, in 1976. She is a full professor for Electrical Systems for Energy, teaching "Electrical Power Systems for Transportation." Her research interests include innovative soft computing and artificial intelligence techniques for power system analysis; the management of electrical loads; reliability and power quality in railway and metrotransit systems; smart grid. Prof. Lamedica is a member of the Italian Ministry of Infrastructure and Transport Technical Committees for transport systems safety and for air and land funicular systems.

**Alessandro Ruvio** received an M.S. degree in Electrical Engineering with honors, awarded with the title "Excellent Graduate", and a Ph.D. degree at the Sapienza University of Rome in 2012 and 2019, respectively. He was an assistant researcher at the Engineering Department of Astronautics, Electrical and Energetic at the Sapienza University of Rome, from 2013 to 2015. In 2019, he became an assistant professor. Holder of two patents. His main research topics have been: energy saving in electrical systems for the transportation of both DC and AC. He is a member of the Italian Federation of Electrotechnics, Electronics, Automation, Informatics and Telecommunications (AEIT); assistant on projects financed by Enea, Navy, and the Ministry of Economic Development; as well as a reviewer for many technical journals. He is also an editorial board member of *Electronics*.

# Preface to "Modeling and Simulation of Electricity Systems for Transport and Energy Storage"

The extensive evolution of electrical systems, due to the increase in distributed generation and renewable sources, has also had effects on electrical transportation systems, in order to improve energy efficiency and environmental sustainability.

In the urban context, the development of not only constrained guideway systems but also those of recent diffusion with hybrid and/or all-electric propulsion such as cars, bikes and scooters, have been characterized by both shorter times and high-power charging requests, with a strong impact on the power systems in terms of stability, PQ, power flows management, etc.

In the extra-urban context, significant technological evolution took place not only in railway transportation systems, but also in the transport of goods by road, with possible highway electrification.

In these new frameworks, energy storage systems are widely used both for increasing energy efficiency and for voltage regulation.

The new electric scenario for transportation systems therefore requires preliminary studies that involve the use of models and calculation procedures suitable for carrying out in-depth analysis.

This book comprises five peer-reviewed articles covering original research articles on the modeling and simulation of electricity systems for transport and energy storage. Topics include optimal siting and sizing methodology to design an energy storage system (ESS) for railway lines, technical–economic comparison between a 3 kV DC railway and the use of trains with on-board storage systems, how to improve electrical feeding substations, by changing transformer technology and by installing dedicated high-power oriented storage systems, algorithms applied to vehicle-to-grid (V2G) technology, thermal investigation and the optimization of an air-cooled lithium-ion battery pack.

**Regina Lamedica, Alessandro Ruvio**
*Editors*

*Article*

# A Thermal Investigation and Optimization of an Air-Cooled Lithium-Ion Battery Pack

**Xiongbin Peng [1], Xujian Cui [1], Xiangping Liao [2,*] and Akhil Garg [3]**

[1]  Key Laboratory of Intelligent Manufacturing, Ministry of Education, Shantou University, Shantou 515063, China; xbpeng@stu.edu.cn (X.P.); 18xjcui@stu.edu.cn (X.C.)

[2]  College of Mechanical Engineering, Hunan University of Humanities, Science and Technology, Loudi 417000, China

[3]  State Key Lab of Digital Manufacturing Equipment & Technology, School of Mechanical Science and Engineering, Huazhong University of Science and Technology, Wuhan 430074, China; akhil@stu.edu.cn

*  Correspondence: 520joff@163.com

Received: 6 May 2020; Accepted: 5 June 2020; Published: 9 June 2020

**Abstract:** An effective battery thermal management system (BTMS) is essential to ensure that the battery pack operates within the normal temperature range, especially for multi-cell batteries. This paper studied the optimal configuration of an air-cooling (AC) system for a cylindrical battery pack. The thermal parameters of the single battery were measured experimentally. The heat dissipation performance of a single battery was analyzed and compared with the simulation results. The experimental and simulation results were in good agreement, which proves the validity of the computational fluid dynamics (CFD) model. Various schemes with different battery arrangements, different positions of the inlet and outlet of the cooling system and the number of inlets and outlets were compared. The results showed that an arrangement that uses a small length-width ratio is more conducive to promoting the performance of the cooling system. The inlet and outlet configuration of the cooling system, which facilitates fluid flow over most of the battery pack over shorter distances is more beneficial to battery thermal management. The configuration of a large number of inlets and outlets can facilitate more flexible adjustment of the fluid flow state and can slow down battery heating to a greater extent.

**Keywords:** thermal management system; optimal configuration; air-cooling; lithium-ion battery

---

## 1. Introduction

The performance of the lithium-ion battery (LIB) is at the core of the driving system of an electric vehicle, and thus it significantly affects the driving range and service life of electric vehicles [1,2]. Among the many factors that determine battery performance, the influence of temperature on the battery should not be underestimated [3]. Battery heating is an inevitable phenomenon and a complicated problem [4]. Reasonable control of temperature changes in battery operation depends on an efficient battery thermal management system (BTMS) [5,6]. Essentially, batteries work to convert chemical energy into electrical energy for the machine to use, so the heat generated is waste energy, which reduces the energy conversion efficiency of the battery. A certain amount of heat yield is beneficial to ensure the normal chemical reaction temperature of the battery and promote the charging and discharging process of the battery. However, if an abnormal amount of heat is generated or the heat does not dissipate in time, the battery suffers overheating, which may potentially cause the active material on the electrodes to peel off and promote electrolyte degradation, that is, cause harm to the battery itself [7–9]. These changes are irreversible and can cause permanent damage to the battery. In addition, because the battery is hot, there are safety risks. An overheated battery can short-circuit due to damage to the internal materials, such as the diaphragm, which sets off a chain reaction that can lead to a fire or even an explosion [10]. Without effective control and management, the harmful effects of a single battery can spread throughout the entire battery pack, and thermal runaway will be amplified, causing the battery pack to be out of control. Therefore, it is very necessary for electric vehicles to design a practical and efficient BTMS [11]. Excellent thermal management benefits the performance of the battery pack [12].

There are three main types of LIBs used in electric vehicles (EVs): (1) prismatic; (2) pouch; and (3) cylindrical batteries. Figure 1 shows these three types of batteries. The prismatic battery can be designed according to the needs of customers. This makes the prismatic battery suitable for almost all kinds of electric cars. This high adaptability also results in prismatic batteries that vary in size, nominal voltage, and other parameters, which makes it difficult to form an industry standard. Pouch batteries are manufactured and packed by superposition. Compared with prismatic batteries, the aluminum alloy shell of the pouch battery pack is replaced with a lighter aluminum plastic film package, which improves the energy density of the whole battery pack. However, a big drawback of soft-pack batteries is their poor consistency, which means they need more sophisticated control and monitoring systems. After years of development, the cylindrical battery has obtained a high degree of standardization, which makes it easy to achieve a unified industry standard. In addition, the cylindrical battery has inherent advantages in regard to heat dissipation, and a good heat dissipation space is formed between the cylinders when packing.

**Figure 1.** A prismatic battery, pouch battery and cylindrical battery (from left to right).

Among these types of battery, the cylindrical battery is the most discussed and well-studied and it is the first mass-produced commercial battery., Thus, much of the research has focused on cylindrical batteries. Most of the studies are based on three different ways of heat dissipation: AC, liquid-cooling (LC), and phase-change material (PCM) cooling. Wang detailed the three cooling methods in his research [13]. AC is widely used for its simplicity, low cost and there is no hidden danger of battery damage. Wang verified the effectiveness of forced-air cooling in ensuring the battery pack operates within the normal temperature range (no more than 40 °C) when the discharge of the current rate of a single battery is set at no more than 3C. Similarly, based on the AC method, Mahamud et al. [14] studied the influence of reciprocating airflow in the BTMS of cylindrical battery packs. This method reduced the battery temperature by about 4 °C (72%) and the maximum temperature (MaxT) by about 1.5 °C. LC is a cooling method that uses a liquid material with a large specific heat capacity (SHC), such as water, to flow through the surface of the battery and take away the generated heat. This method requires the battery system to be highly sealed, but its strong heat transfer capacity leads to a better cooling effect than that of AC [15]. Wang et al. [16] proved that several factors, such as fluid flow, flow direction, etc. determine the cooling effect at a certain degree. They designed a BTMS based on a hot silicon plate and used experiments and simulations to explore and verify the significant influence of liquid flow, flow direction and the number of cooling channels on the cooling capacity of the BTMS. PCM cooling, which uses the process of absorbing heat during the phase change to balance the heat that the battery generates, is costly but offers a pollution-free, high-return solution. By comparing cooling performance under different conditions, Kizilel et al. [17] proved that the PCM cooling method is superior to AC in regard to economy, effectiveness, and safety. Through simulation and experiment,

Huang et al. [18] proved that the cooling system of thermal-assisted expanded graphite is superior to the AC system in actual and extreme conditions. In addition, Wang [19] and Wang [20] greatly improved the performance of the cooling system without changing the cooling mode, by optimizing the battery pack structure. Yang et al. [21] found the specific parameters of the battery pack through optimization analysis. They pointed out that the cooling system works best when the height of the battery pack is 34 mm and the width is 32 mm.

In this study, a cylindrical BTMS based on AC is proposed. The battery pack consists of 20 cylindrical battery modules of type 18650 batteries. First, the dynamic model was built and the validity of the model was verified. Then, the BTMS of the cylindrical battery pack was optimized for the comparative analysis of different battery configurations with different cell layouts, different positions of the air inlet and outlet in the AC system and the number of inlets and outlets. Finally, the accuracy and effectiveness of the whole process were verified experimentally, which provided an ideal design for the BTMS of the cylindrical battery in the future.

## 2. Problem Description

The single 18650 LIB used in this paper has a voltage of 3.6–4.2 V and a capacity of 2600 mAh. The main aim of this study is to reduce the MaxT and temperature standard deviation (TSD) of the battery during operation by optimizing the configuration of the BTMS for the battery module, which is composed of 20 cells. For a multi-cell module, the design of the BTMS must cover many aspects. The layout of the battery, the positioning scheme of the inlets and outlets, the number of inlet and outlet as well as the structural size of the battery module all affect the final cooling effect of a BTMS. First, the layout plan of the battery pack was analyzed and designed, then the optimal inlet and outlet position was designed based on the optimized layout plan. Then the influence of the number of inlet and outlet on the thermal performance of the battery was explored by increasing the number of inlets and outlets. Finally, the optimized configuration was determined. Details of the research method and technical route are shown in Figure 2. The specific steps to solve this multi-objective coupling problem are as follows:

(1)  Thermal management system and CFD.
(2)  Optimized configuration of the battery pack scheme.
(3)  Experimental validation and result analysis.

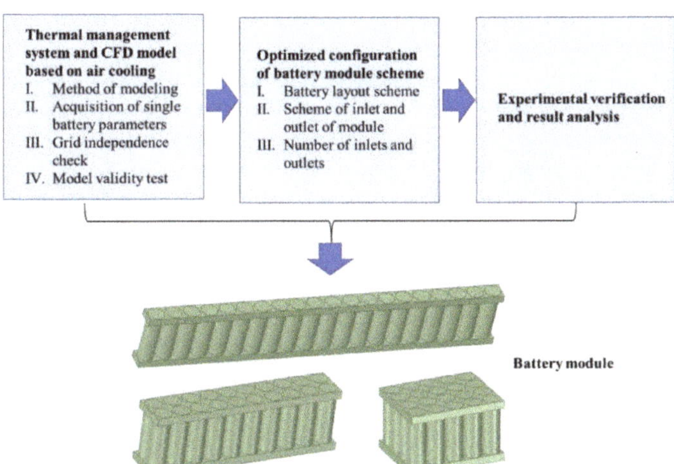


**Figure 2.** Research methods and the technical route.

### 3. Heat Dissipation Model of Computational Fluid Dynamics

There are four main sources of heat production in working batteries, namely, reaction heat (RH), side reaction heat (SRH), joule heat (JH) and polarization heat (PH). The total calculation of the heat generation is shown in Equation (1).

$$P_{total} = P_{re} + P_{sr} + P_{jo} + P_{po} \tag{1}$$

where $P_{total}$ represents the power of the total heat, $P_{re}$ represents the power of RH, $P_{se}$ represents the power of SRH, $P_{jo}$ represents the power of JH, and $P_{po}$ represents the power of PH.

RH refers to the heat generated by the chemical reaction in the electrodes in charging and discharging. Generally, the charging process of LIB absorbs energy to reduce ambient temperature and in reverse, its discharging process releases heat [22]. This results in a side reaction, that is, heat is generated by a series of chemical reactions other than the main chemical reaction, such as the partial decomposition of electrolyte at high temperature and self-discharge caused by the change in electrode material structure. These side reactions are intensified in the period before the battery fails. However, during the life of the battery, the side reactions are so weak that the heat of the side reaction is usually ignored. JH is the work done by the current on the internal resistance (IR). This part of the heat can be calculated by Joule's law as shown in Equation (2):

$$P_{jo} = I^2 R_\Omega \tag{2}$$

where $R_\Omega$ is IR, and $I$ refers to the current on $R_\Omega$.

PH refers to the heat generated when the positive and negative electrode potential deviates from the equilibrium potential. When polarization occurs, the voltage difference between the battery's open circuit voltage and the terminal voltage generate PH. Generally, it is assumed that there is a polarization IR $R_p$, and the heating power is calculated by Joule's law, as shown in Equation (3),

$$P_{po} = I^2 R_p \tag{3}$$

*3.1. Acquisition of Battery's Thermodynamic Parameters*

Before the heat dissipation model is built, the battery parameters need to be determined. These parameters include the IR, SHC, heat yield and thermal conductivity (TC) of the single battery. In order to obtain these parameters, it is necessary to conduct charge and discharge experiments on a single battery. The equipment generates cycles of charge and discharge, and the measurement of the parameters of current, voltage and temperature. The Arbin machine battery testing system was adopted, which can charge/discharge batteries at a set constant current or voltage value and record the respective current, voltage, capacity, and impedance per minute simultaneously. Besides the Arbin machine, several temperature sensors and a miniature blower were needed. The equipment is shown in Figure 3a and each battery is equipped with three temperature sensors, as shown in Figure 3b. For determining the measurement error, every single cell is installed with three temperature sensors that are attached at the cathode, middle and anode of the cell. The temperature utilized in this study is the average value.

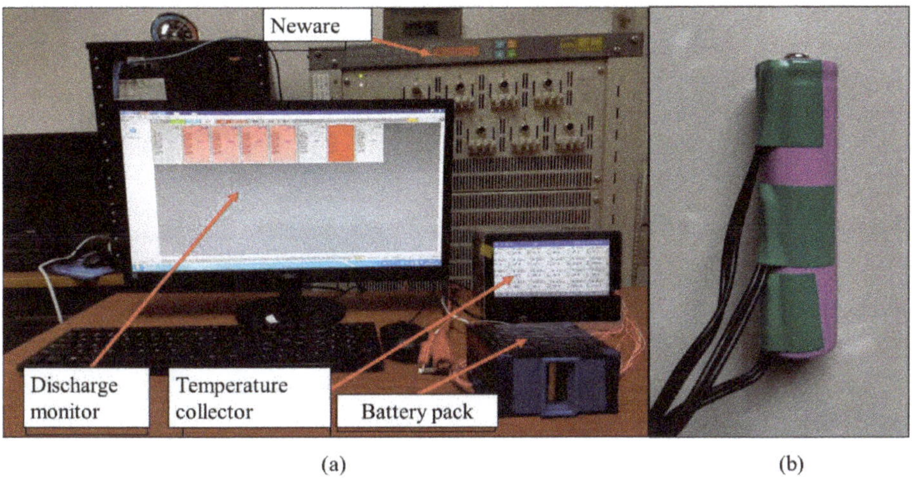


**Figure 3.** (**a**) Experimental equipment. (**b**) The battery with temperature sensors.

### 3.1.1. Internal DC Resistance

The battery resistance refers to the direct current (DC) resistance of the battery, which consists of the static resistance and polarization resistance. The hybrid pulse power characterization test (HPPC) method was used to measure the DC resistance of the battery. The HPPC method is designed to test the resistance of the battery under specific temperatures and state of charge (SOC).

### 3.1.2. Specific Heat Capacity

The type 18650 battery is not homogeneous, so, the SHC $C_M$ refers to the equivalent SHC of a single battery. Assuming that the equivalent SHC of the battery is $C_M$, according to the definition of SHC, it is described by Equation (4):

$$\frac{d\phi}{dt} = C_M M \frac{dT}{dt} \tag{4}$$

It is known that the value of the heat generated (i.e., temperature rise) by the battery is directly proportional to the equivalent SHC of the battery. In fact, the SHC is computed by the law of conservation of energy (COE). In an adiabatic environment, the value of the heat generation of a battery equals its stored heat. The accumulation of heat storage will lead to the rise in battery temperature. By measuring the temperature rise in a specific period time, the SHC of the battery can be calculated according to the above formula. In the actual experiment, in order to reduce the experimental error, the SHC of the battery under different discharge ratios (i.e., different value of heat generation) was measured and the SHC of the battery was calculated by linear regression. The experimental procedures have been described in detail in [23].

### 3.1.3. Heat Generation of Single-Cell and Thermal Conductivity

On this basis, Bernardi et al. (1984) provided a methodology to calculate the heat production rate of batteries, which is expressed by Equation (5).

$$P_{\text{total}} = -IT\frac{dE}{dT} + I(E - U) \tag{5}$$

where $P$ is the power of the heat generation, $I$ is current in circuit, $T$ is the real-time temperature of the battery, $E$ refers to the open circuit voltage of battery, and $U$ refers to the average terminal voltage of battery.

In Equation (5), the first item $IT\frac{dE}{dT}$ on the right side of equation is the formula to calculate the power of RH, which equals the $P_{\text{total}}$ in Equation (1). The second item refers to summary of JH and PH and it shows the voltage attribution. The voltage drops $(E - U)$ in the open circuit voltage and terminal voltage are attributed to internal DC resistance and their relationship is represented by Equation (6).

$$I(E - U) = I^2(R_\Omega + R_P) \tag{6}$$

$R_\Omega + R_P$ make up the internal DC resistance, which can be expressed by $R$. So, the heat production Equation (5) can be expressed by Equation (7),

$$P_{\text{total}} = -IT\frac{dE}{dT} + I^2R \tag{7}$$

On the premise that the equivalent SHC is known (calculated in Section 3.1.2), the heat production in a certain period of time can be solved according to the definition of the SHC [23]. In this experiment, the heat dissipation effect of the battery within 15 min is discussed, so the heat generation of the battery should also be the heat generation within 15 min. The discharging current rate is set at 2C. Twenty single batteries were selected for heat yield measurement. The specific operational steps are as follows:

A.  Take 20 type 18650 batteries and fully discharge them, and then hold them for 30 min at room temperature before charging them at 0.75C.
B.  Record the current, voltage, temperature and corresponding time value.

For each cell, the temperatures were recorded. During battery discharge, oxidation occurs in the negative electrode, and lithium is separated from the carbon rod and releases energy, which causes the temperature of the negative electrode to rise. At the same time, there is a reduction at the positive pole. Lithium ions precipitate at the positive electrode and absorb a certain amount of energy, so the temperature of the positive pole decreases. Therefore, during the whole discharge process, the temperature gradually decreases from the negative electrode to the positive electrode. Thus, it is assumed that the temperature of the positive pole is transferred from the negative electrode and the equivalent TC of the whole battery is calculated based on the battery temperature distribution. The formula to calculate TC is illustrated in Equation (8):

$$\lambda = \frac{q\delta}{\Delta t} \tag{8}$$

where $\lambda$ is TC, $q$ is conducted heat, and $\delta$ is the distance from the negative pole to the positive pole. $\Delta t$ refers to the temperature difference (TD) between the two poles. In fact, the conductivity of the battery is anisotropic, which means the TC is different in the surface and the thickness directions. However, the TC in the surface direction is much less than that in the thickness direction and it is ignored in this model [23,24].

In order to avoid contingency and remove outliers, 20 type 18650 batteries were tested under the same conditions. After calculating the average value of the 20 batteries, the thermodynamic parameters are listed in Table 1.

**Table 1.** The thermodynamic parameters of the battery.

| Parameter | $\rho$ (kg·m$^{-3}$) | $R_{DC}$ ($\Omega$) | $Cp$ (J·kg$^{-1}$·K$^{-1}$) | $\lambda$ (W·m$^{-1}$·K$^{-1}$) | $P$ (W/m$^3$) |
|---|---|---|---|---|---|
| Value | 2812.7 | 0.074 | 922.4 | 141.2 | 14,947.6 |

### 3.2. Modeling Method

A three-dimensional model was built in SpaceClaim18.2. The meshing operation and numerical simulation were processed in ansys18.2. In order not to consider the energy consumption too much, the fluid inlet speed is set to 1 m/s and it is maintained by a fan in all designs. This is the typical forced convection mode. The method of forced convection improves the heat dissipation effect by increasing the heat transfer rate per unit time. Compared with natural convection, it consumes more energy but performs very well with respect to thermal management. The Reynolds number can be calculated by Equation (9). In these models, the distance between the battery and the wall is 10 mm. The gap in the battery center is 20 mm. The fluid cross section with the largest characteristic length exists in a rectangular plane of 2 × 490 mm. The max Reynolds number here is 406, which determines the laminar flow type in the simulation.

$$\mathrm{Re} = \frac{\rho v d}{\mu} \tag{9}$$

where $\rho$ and $\mu$ are fluid density and dynamic viscosity coefficient, and $v$, $d$ are the characteristic velocities and characteristic length of the flow field.

The results of the grid independence test are shown in Figure 4, which ensure that the subsequent simulation process is reliable. In Figure 4, once the number of grids reaches 359,370–389,048, the MaxT of the model remains constant, that is, when the number of grids is more than 359,370, the size of the nodes will not affect the calculation results and the grid will pass the independence test. In order to save computing costs, the number of grids is selected as 359,370.

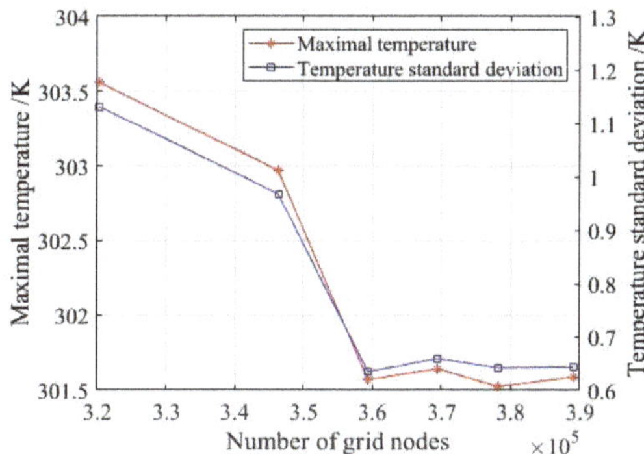


**Figure 4.** Grid independence test based on the maximum temperature (MaxT) and temperature standard deviation (TSD).

To simulate the flow model in ANSYS and compare the results, some assumptions need to made, including:

1.  Steady-state flow from the inlet
2.  Constant thermal and physical properties
3.  No energy within flow
4.  Heat loss of radiation and natural convection is neglected
5.  Gravity is not taken into consideration.

In the process of battery heat generation, the energy of the battery conforms to the law of COE [25,26], that is, it satisfies Equation (10),

$$\rho C_p \frac{\partial T}{\partial t} = \frac{\partial}{\partial x_d}(k_{xd}\frac{\partial T}{\partial x_d}) + \frac{\partial}{\partial y_d}(k_{yd}\frac{\partial T}{\partial y_d}) + \frac{\partial}{\partial z_d}(k_{zd}\frac{\partial T}{\partial z_d}) + Q_v \tag{10}$$

where $\rho$ refers to density of battery; $C_p$ is equivalent to SHC; $T$ refers to the battery temperature; $k_{xd}$, $k_{yd}$, and $k_{zd}$ represent the heat conductivity coefficient (HCC) in the $x$, $y$, and $z$ directions, respectively; and $Q_v$ is a volumetric heat source of battery, which is the same as the heat generation rate.

The energy conservation equation (ECE) for coolant [8] is expressed by Equation (11):

$$\rho_{co}\frac{\partial T_{co}}{\partial t} + \nabla(\rho_{co}\vec{v}T_{co}) = \nabla(\frac{k_{co}}{C_{co}}\nabla T_{co}) \tag{11}$$

The coolant in this study is air, so the variable $\rho_{co}$ refers to the density of air; $C_{co}$ and $k_{co}$ are the SHC and TC of the air, respectively.

The velocity of air is about 1 m/s whose Mach number is far less than 0.3. So, the continuous equation of coolant (which here refers to air) is as shown in Equation (12)

$$\nabla\vec{v} = 0 \tag{12}$$

where $\vec{v}$ is the velocity vector.

The momentum conservation equation (MCE) is shown in Equation (13).

$$\rho_{co}\frac{d\vec{v}}{dt} = -\nabla P + \mu\nabla^2\vec{v} \tag{13}$$

where $P$ and $\mu$ are the static pressure and dynamic viscosity of the air, respectively.

### 3.3. Model Validity Test

To ensure that the model has practical value, it is necessary to test the validity of the model. A validity test involved a pre-experiment and a pre-experimental simulation and then the results are compared to show whether the model is effective and can be used for further study. A single battery was selected for testing and it was covered by a piece of insulated cotton. The covered single battery used to simulate adiabatic condition is shown in Figure 5a. Then, the battery was continuously discharged to collect the temperature change data for the battery during discharge. The result is shown in Figure 5b. The temperature measured by the experiment was almost the same as that measured by the simulation.

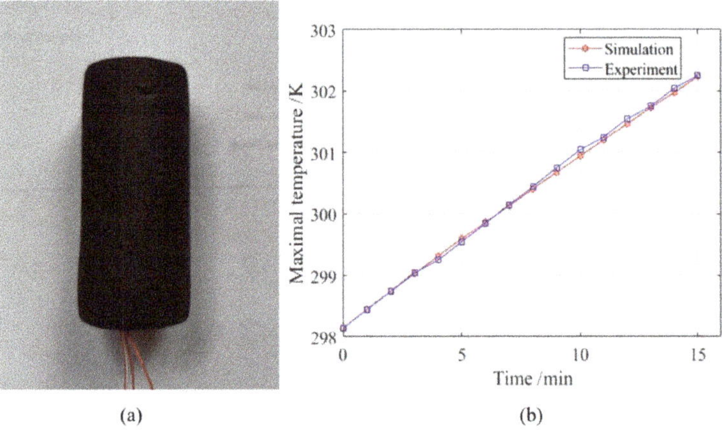


(a)                              (b)

**Figure 5.** (a) The covered single battery. (b) Temperature rise of a single battery under an adiabatic environment.

## 4. Analysis of the Results

The temperatures based on the battery layout, the inlet and outlet position, and the number of inlets and outlets of the module are considered. An AC system was adopted as the cooling method due to its simplicity and low cost. The forced air speed was set at 1 m/s. and a blower provides the forced air that flows into the battery pack. The accurate measurement of the anemometer ensures that the wind speed of the battery pack inlet is constant at the desired value. The blower and anemometer used in this study are shown in Figure 6. The discharge time was 15 min. Temperature was recorded at the end of each minute. The initial distance between the battery center and the wall is 20 mm, and the shortest distance between the battery center and the wall is 10 mm. Battery material parameters are shown in Table 1.

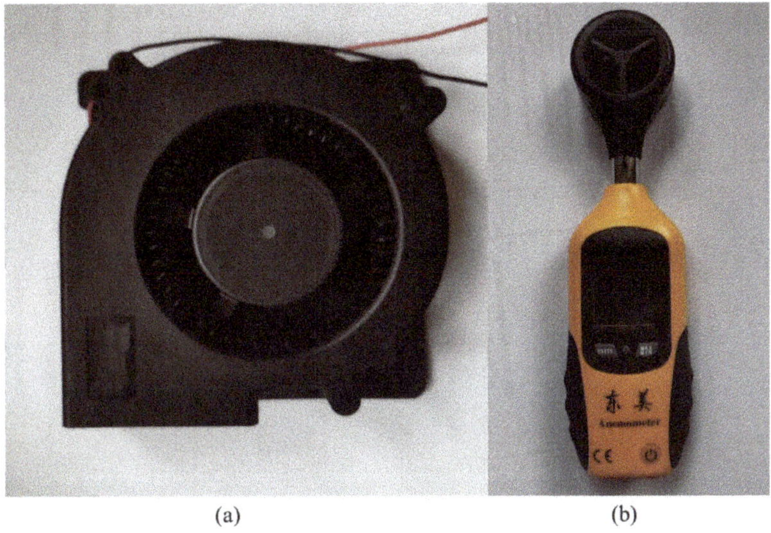


**Figure 6.** The blower (**a**) and anemometer (**b**) used in experiment.

*4.1. Battery Layout Analysis*

The battery pack contains 20 type 18650 batteries. There are three possible layout configurations for a rectangular layout: $1 \times 20$, $2 \times 10$ and $4 \times 5$. The layouts are shown in Figure 7. To choose the best scheme, the three schemes need to be simulated and predicted.

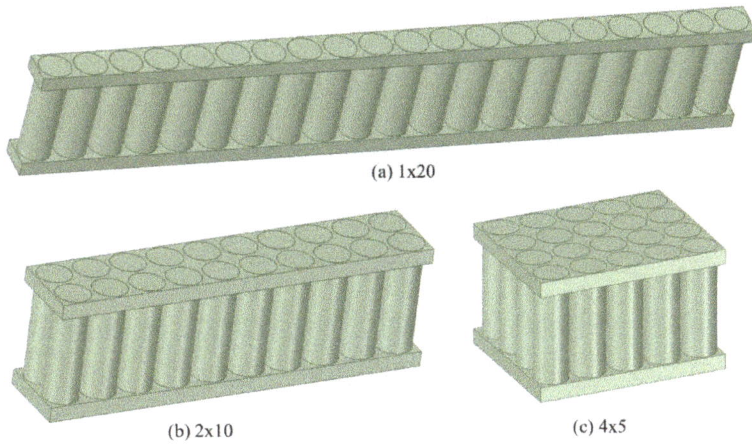


**Figure 7.** Battery layout configuration (**a**) $1 \times 20$, (**b**) $2 \times 10$, (**c**) $4 \times 5$.

To fully compare the advantages and disadvantages of these three configurations and select the best battery configuration, a single variable control method was used to make a comparative analysis of the schemes. For each arrangement, there is an air inlet and an air outlet equal in size to a rectangle of 20 × 49 mm. The module is cooled with constant wind speed in both the transverse (h) and longitudinal (v) directions. The six temperature distributions of the three battery configurations are shown in Figure 8. During the 15-min discharge, the MaxT variation of the layout schemes at the end of each minute is shown in Figure 9.

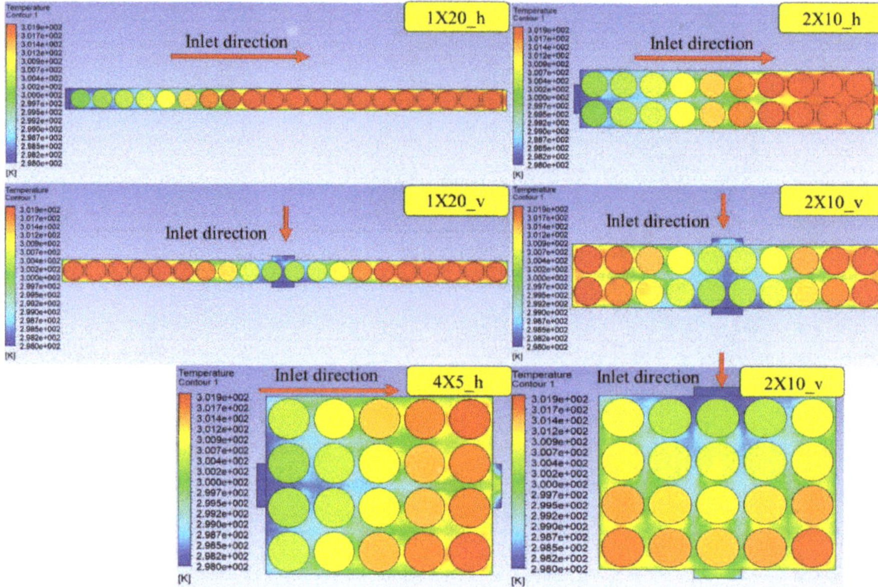


**Figure 8.** Temperature distribution of the battery layout schemes.

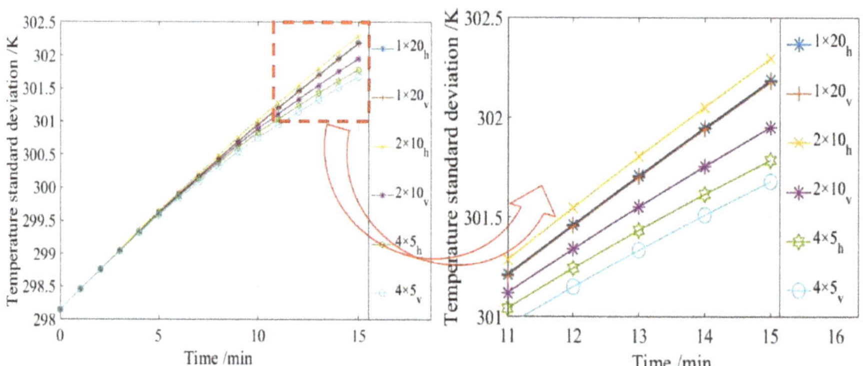


**Figure 9.** The MaxT variation in the layout schemes over time.

In Figure 8, it can be seen that the low-temperature area appears at the entrance and the high-temperature area is distributed at the exit. This is in line with the actual situation, because the air temperature entering the inlet is room temperature (298.15 k). When the air flows through the battery surface, which has a higher temperature, it will absorb the heat of the battery leading to a gradual increase in the temperature of air from the inlet to the outlet. Additionally, the longer the air travels through the cooling system, the hotter it gets. As seen in Figure 9, the cooling effect of the $4 \times 5$ battery arrangement is better in both directions than the other two arrangements. Therefore, the battery layout configuration of $4 \times 5$ was selected as the battery layout.

### 4.2. Inlet and Outlet Position

As mentioned above, the optimal battery layout configuration is a $4 \times 5$ scheme. The four sides of the battery pack can be set as an air inlet and air outlet. With each side divided into left, center and right parts, all 12 parts can be used as inlet or outlet, as shown in Figure 10.

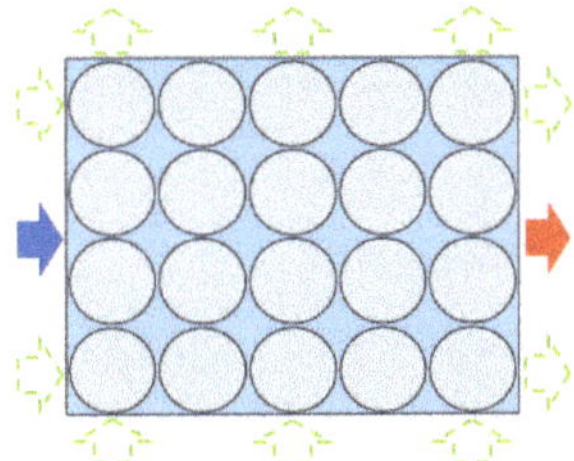

**Figure 10.** Optional choices of inlets and outlets.

According to symmetry, the duplicate and inferior schemes were removed, leaving 18 schemes to choose from. These 18 schemes are shown in Figure 11.

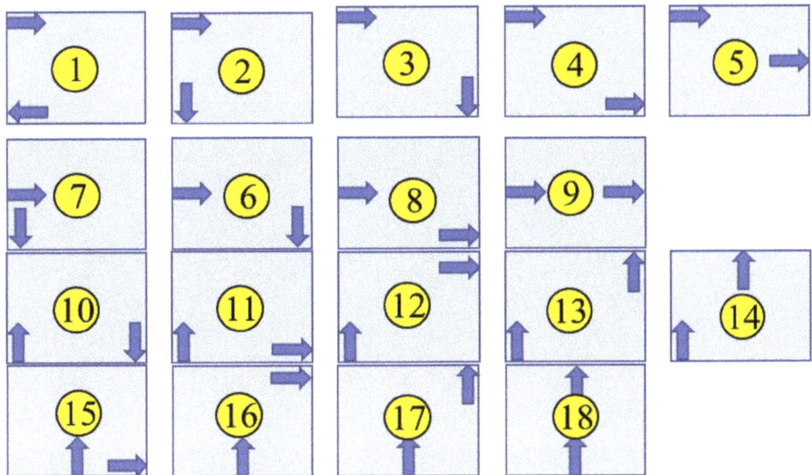

**Figure 11.** Schemes for the battery pack with a single inlet and single outlet.

The MaxT change curve and TSD change values of all schemes were extracted, as shown in Figures 12 and 13. The temperature of all schemes equals room temperature (298.15k) at the initial moment. With the discharge process, the MaxT increases gradually and the rate of increase gradually slows down. It can be seen that the final MaxT of scheme 1, 2, 7, and 15 is relatively high. The final MaxT of scheme 3, 4, 12, and 13 is relatively low. In addition, the air speed at the inlet and outlet are listed in Figure 14. The speed at the inlet remained at 1 m/s while it was no more than 0.2 m/s at the outlet. The outlet air speed of scheme 6 is the lowest, while the outlet air speed is the highest in scheme 3. The difference in outlet air speed indicates the complexity of flow. The eight schemes for the fluid velocity distribution path are shown in Figures 15 and 16. It can be seen that in schemes 1, 2, 7 and 15, the inlet and outlet are too close and the pressure difference between the inlet and the outlet is the largest. After entering the inlet of the battery pack, the air fails to flow through most areas of the battery bag, resulting in most of the heat not being taken away in time. Therefore, the heat dissipation effect of the battery pack is poor. Scheme 3, 4, 12, and 13 avoid this problem. The fluid flows through almost all the cells in the battery pack.

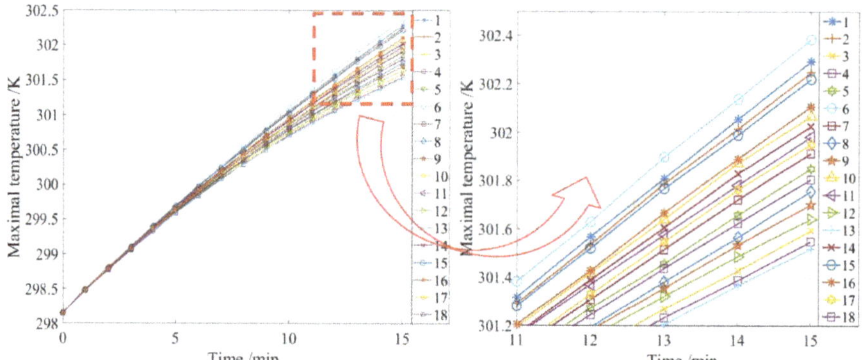

**Figure 12.** The MaxT variation in schemes for the battery pack with a single inlet and single outlet over time.

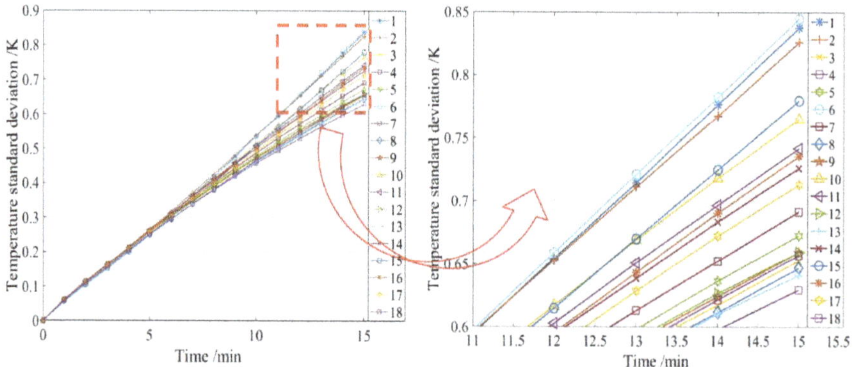

**Figure 13.** TSD of schemes for the battery pack with a single inlet and single outlet over time.

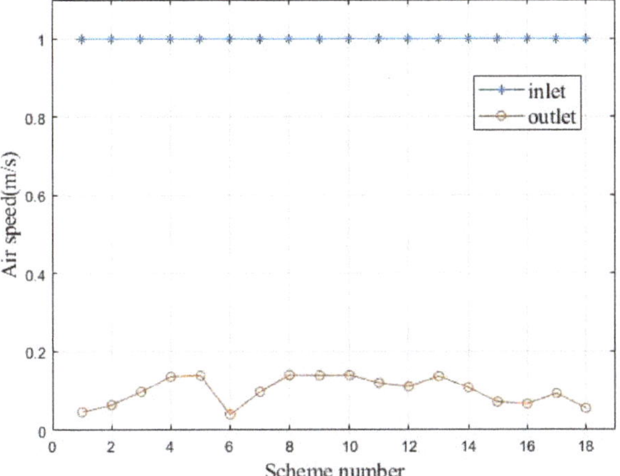

**Figure 14.** The air speed at the inlet and outlet based on inlet and outlet position.

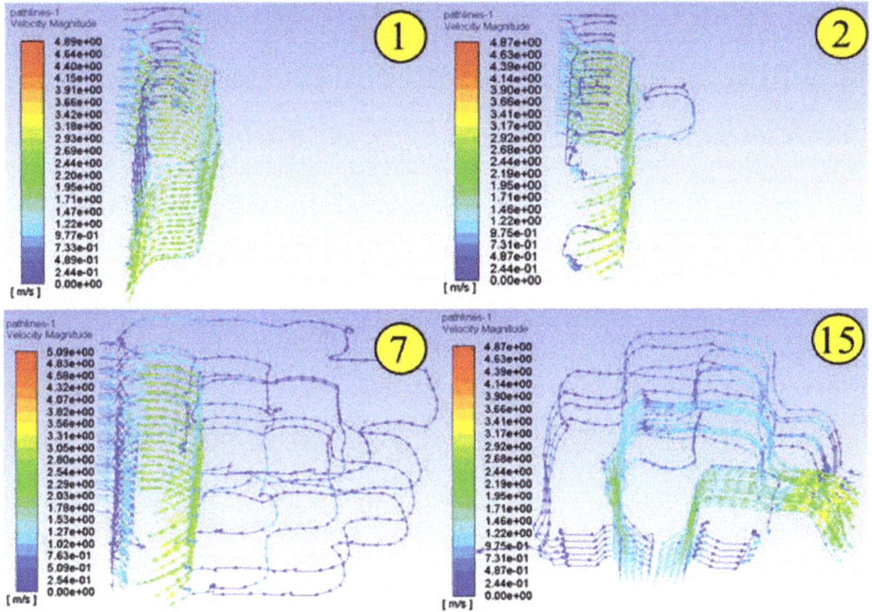

**Figure 15.** Velocity path line of scheme 1, 2, 7, 15.

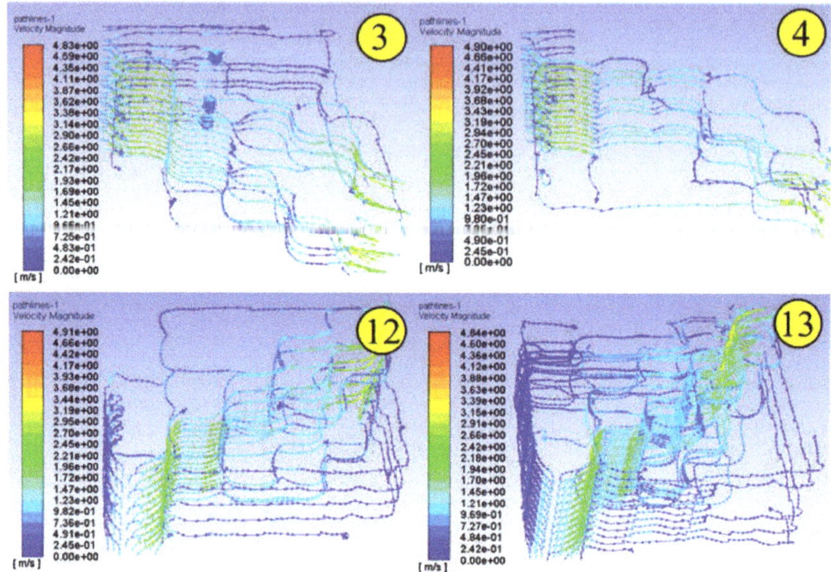

**Figure 16.** Velocity path line of scheme 3, 4, 12, 13.

### 4.3. Number of Inlets and Outlets

In order to obtain a better BTMS, a model with multiple outlet modes was studied. Options include one inlet and two outlets, two inlets and one outlet, two inlets and two outlets, and so on. Eighteen one inlet and one outlet options were discussed above in Section 4.2. On the basis of these one inlet and one outlet scheme, the one inlet and two outlet configurations are further discussed. Considering the symmetry, the inlet is arranged in the middle the side. The specific scheme is shown in Figure 17.

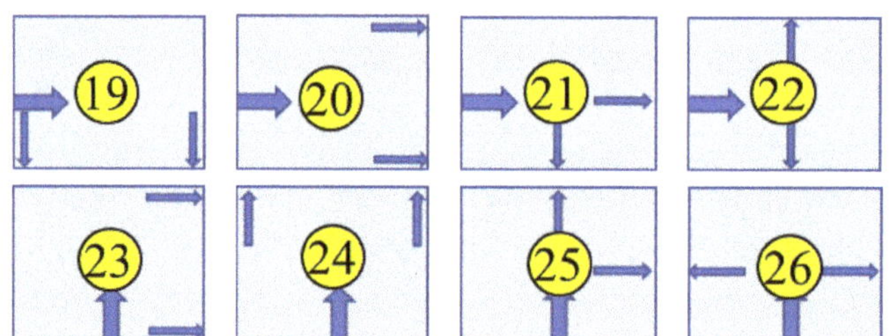

**Figure 17.** Schemes for the battery pack with a single inlet and double outlet.

According to the information shown in Figures 18 and 19, the MaxT is in scheme 23 and the minimum is in scheme 24. The largest TSD is in scheme 19 and the smallest is in scheme 24. As shown in Figure 20 by comparing scheme 19 and scheme 23, the short flow path of the fluid will remove local heat, and the temperature of the whole battery pack will be higher. The longer the fluid flow path, the more uniform the temperature distribution of the battery pack and the smaller the TSD. The MaxT and the TSD of different inlet and outlet schemes are caused by the change in the fluid flow state. The air speed at the inlet and outlet is displayed in Figure 21. The speed at the inlet was 1 m/s. There were discrepancies in the outlet speed. The zigzag change manifested by several schemes were at similar outlet speeds.

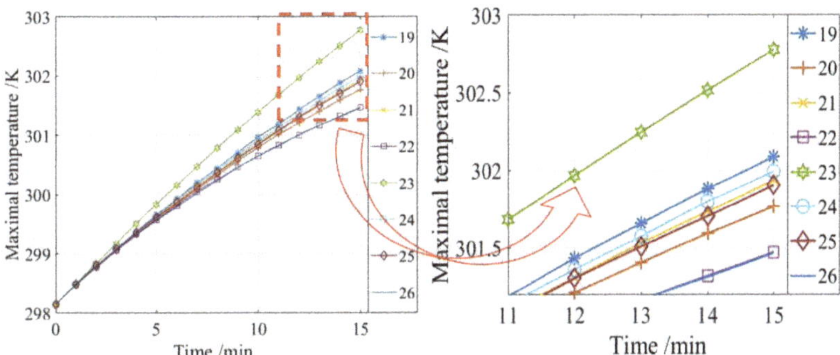

**Figure 18.** The maximal temperature of schemes for the battery pack with a single inlet and single outlet over time.

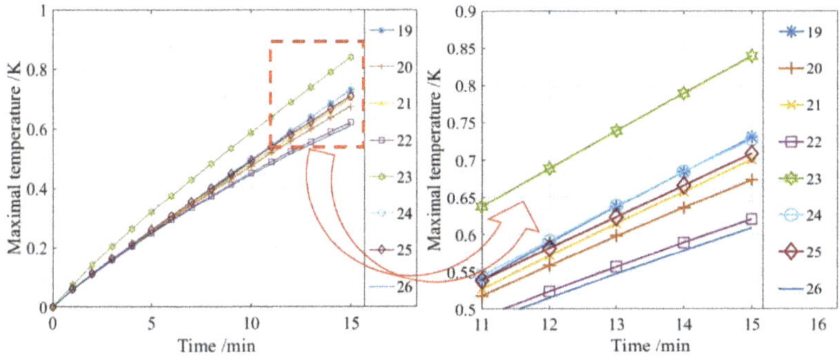

**Figure 19.** TSD of schemes for the battery pack with a single inlet and single outlet over time.

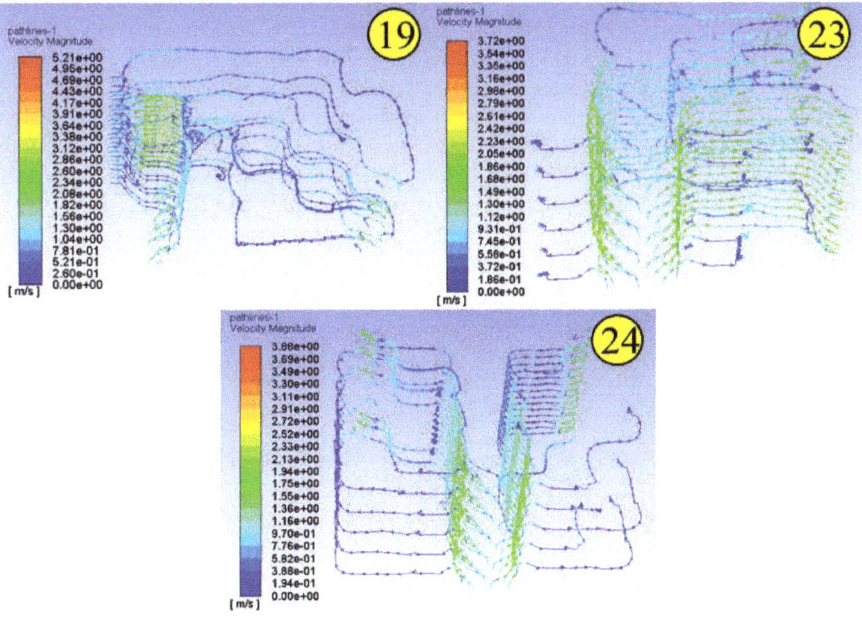


**Figure 20.** Velocity path line of scheme 19, 23, 24.

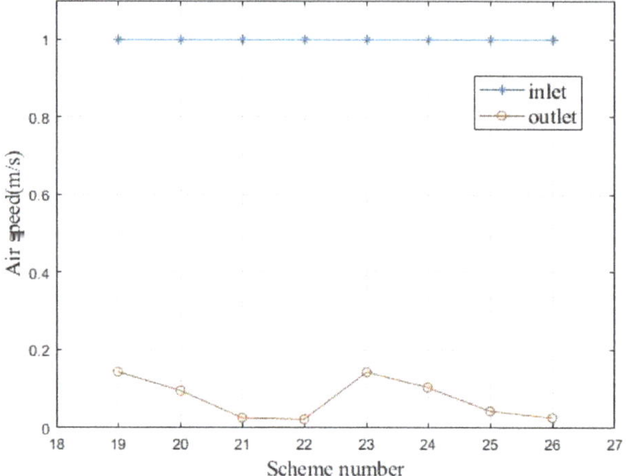

**Figure 21.** The air speed at the inlet and outlet based on the number of inlets and outlets.

## 5. Experimental Validation

As we know from the previous section, the optimal arrangement of the inlet and outlet in the one-in-one-out model is scheme 13, and the optimal arrangement of the inlet and outlet in the one-in-two-out mode is scheme 24. This is because, under these two configurations, the fluid flow state in the battery pack is most suitable for dissipating the heat in the battery pack, which can result in good heat dissipation of the battery system when working. However, the simulation results need to be verified by experiments. Therefore, discharge tests were conducted on scheme 13 and scheme 24. The 20 batteries are arranged in a 4 × 5 arrangement and connected in a series of four and a parallel of five. To ensure a constant discharge current of 2 A, the discharge current of 8 A was set to constantly discharge for 15 min. In the experiment, the temperature of 20 batteries was collected in real-time, and a temperature sensor was installed on the side of each battery. The experimental set up is shown in Figure 3.

The temperature data of two battery packs were collected in the experiment. The temperature at the end of discharge (i.e., the MaxT of the battery) was marked on the single battery according to the battery layout in the experiment. The results are shown in Figures 22 and 23, respectively. Under the premise of good enough performance of the BTMS, lower cost should also be a focus. The transient change process in the temperature cloud map is shown in Figures 24 and 25.

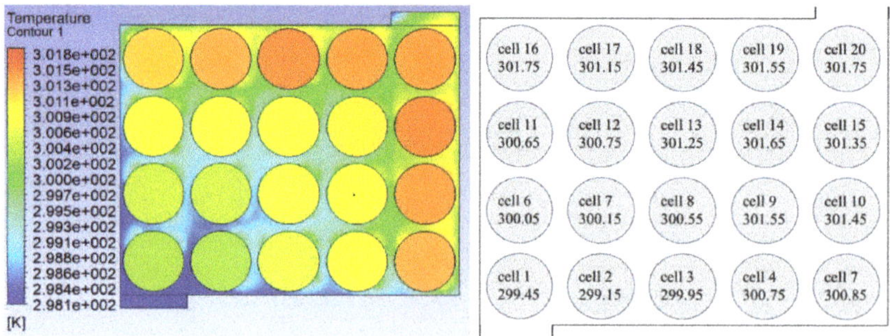


**Figure 22.** Comparison between the simulation and experimental results for scheme 13 (simulation on the left and experiment on the right).

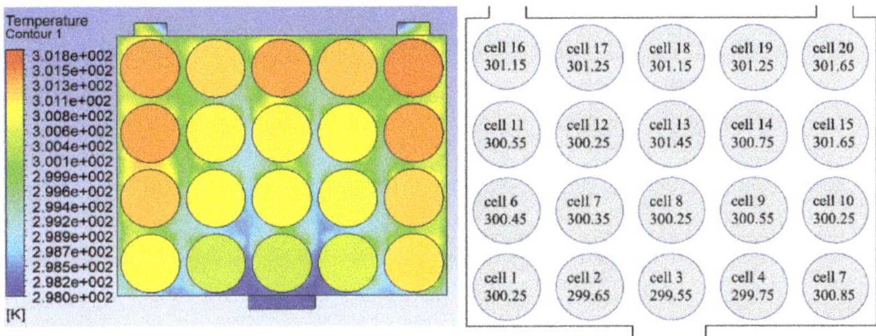

**Figure 23.** Comparison between the simulation and experimental results for scheme 24 (simulation on the left and experiment on the right).

Figures 22 and 23 show the comparison of the battery temperature during the simulation and the experiment. In Figure 22, the error in MaxT between the simulation and experiment is 0.017% and in Figure 23, the error in MaxT between the simulation and experiment is 0.049%. The experimental results are close to the simulation temperature. Battery temperature is generally lower near the inlet, and the temperature starts to rise as the distance from the inlet increases. High-temperatures appear near the outlet, and the closer they are to the outlet, the higher the temperature. These phenomena are consistent with the basic predictions of heat transfer. However, the experimental results are still inconsistent with the simulation results. At the end of discharge, the battery temperature is still slightly lower than the simulation results. In addition, the influence of pressure on the temperature field was not considered in this study, which is a possible cause of temperature error. Table 2 shows a comparison of the air speed in the simulation and the experiment. Limited by the difficulty of adjustment and measurement accuracy, the forced air speed at the inlet can only be adjusted close to 1 m/s. The error in the simulation and experiment stabilized at less than 15.38%.

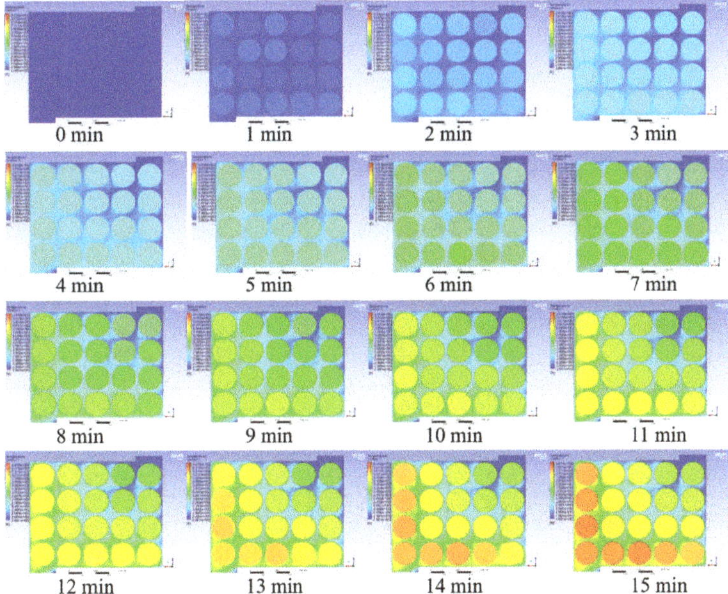


**Figure 24.** A temperature field change in scheme 13 in the dying minutes.

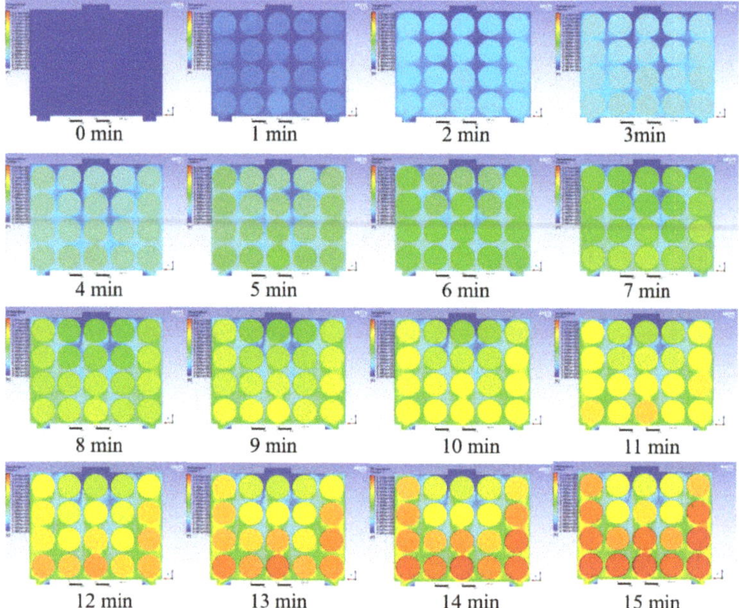


**Figure 25.** A temperature field change in scheme 24 in the dying minutes.

**Table 2.** Comparison of the air speed at the inlet and outlet in the simulation and the experiment.

|  | Position | Simulation | Experiment |
|---|---|---|---|
| Scheme 13 | Air speed at inlet | 1 m/s | 1.03 m/s |
|  | Air speed at outlet | 0.136 m/s | 0.15 m/s |
| Scheme 24 | Air speed at inlet | 1 m/s | 1.01 m/s |
|  | Air speed at outlet | 0.026 m/s | 0.03 m/s |

## 6. Conclusions

In this study, we designed an AC system for a cylindrical battery pack. The experiment was conducted on the basis of the simulation, and the following conclusions were drawn by comparing the results:

(1) For the rectangular arrangement of a multi-cell cylindrical battery pack, an arrangement with a small length-width ratio is more conducive to reducing the MaxT and TSD of the battery in the discharge process.
(2) The inlet and outlet configuration of the cooling system that facilitates fluid flow over most of the battery pack over shorter distances, is more conducive to battery thermal management.
(3) The configuration of a large number of inlets and outlets facilitates more flexible adjustment of the fluid flow state and can slow down battery heating to a greater extent.

As the distance of cells in the pack greatly determines the performance of thermal management, future work will consider the optimization of the structural parameters.

**Author Contributions:** Data curation, X.C.; Formal analysis, X.P. and X.C.; Funding acquisition, X.L., X.P. and A.G.; Methodology, X.P. and X.L.; Software, A.G. All authors have read and agreed to the published version of the manuscript.

**Funding:** This was funded by the Department of education of Guangdong Province of China, grant number 2019KTSCX039; the Natural Science Foundation of Hunan Province of China, grant number 2017JJ3057; and the Research Foundation of Education Bureau of Hunan Province, China, grant number 17C0473. The APC was funded by Shantou University.

**Conflicts of Interest:** The authors declare no conflict of interest.

## Nomeclature

| Abbreviation | Meaning |
| --- | --- |
| LIB | Lithium-ion battery |
| BTMS | Battery thermal management system |
| AC | Air-cooling |
| LC | Liquid-cooling |
| PCM | Phase change material |
| RH | Reaction heat |
| SRH | Side reaction heat |
| JH | Joule heat |
| PH | Polarization heat |
| IR | Internal resistance |
| SHC | Specific heat capacity |
| TC | Thermal conductivity |
| SOC | State of charge |
| HCC | Heat conductivity coefficient |
| COE | Conservation of energy |
| TD | Temperature difference |
| MaxT | Maximum temperature |
| MinT | Minimum temperature |
| TSD | Temperature standard deviation |
| ECE | Energy conservation equation |
| MCE | Momentum conservation equation |
| CFD | Computational fluid dynamics |

## References

1. Mohammed, A.H.; Esmaeeli, R.; Aliniagerdroudbari, H.; Alhadri, M. Dual-purpose cooling plate for thermal management of prismatic lithium- ion batteries during normal operation and thermal runaway. *Appl. Therm. Eng.* **2019**, *160*, 114106. [CrossRef]
2. Song, L.; Liang, T.; Lu, L.; Ouyang, M. Lithium-ion battery pack equalization based on charging voltage curves. *Int. J. Electr. Power Energy Syst.* **2020**, *115*, 105516. [CrossRef]
3. Xu, X.; Li, W.; Xu, B.; Qin, J. Numerical study on a water cooling system for prismatic LiFePO 4 batteries at abused operating conditions. *Appl. Energy* **2019**, *250*, 404–412. [CrossRef]
4. Wang, Q.; Jiang, B.; Xue, Q.F.; Sun, H.L.; Li, B.; Zou, H.M.; Yan, Y.Y. Experimental investigation on EV battery cooling and heating by heat pipes. *Appl. Therm. Eng.* **2015**, *88*, 54–60. [CrossRef]
5. Jilte, R.D.; Kumar, R.; Ahmadi, M.H. Cooling performance of nanofluid submerged vs. nanofluid circulated BTMSs. *J. Clean. Prod.* **2019**, *240*, 118131. [CrossRef]
6. Lai, Y.; Wu, W.; Chen, K.; Wang, S.; Xin, C. A compact and lightweight liquid-cooled thermal management solution for cylindrical lithium-ion power battery pack. *Int. J. Heat Mass Transf.* **2019**, *144*, 118581. [CrossRef]
7. Weng, J.; Ouyang, D.; Yang, X.; Chen, M.; Zhang, G.; Wang, J. Alleviation of thermal runaway propagation in thermal management modules using aerogel felt coupled with flame-retarded phase change material. *Energy Convers. Manag.* **2019**, *200*, 112071. [CrossRef]

8.   Liu, W.; Jia, Z.; Luo, Y.; Xie, W.; Deng, T. Experimental investigation on thermal management of cylindrical Li-ion battery pack based on vapor chamber combined with fin structure. *Appl. Therm. Eng.* **2019**, *162*, 114272. [CrossRef]

9.   Liu, K.; Li, K.; Ma, H.; Zhang, J.; Peng, Q. Multi-objective optimization of charging patterns for LIB management. *Energy Convers. Manag.* **2018**, *159*, 151–162. [CrossRef]

10.   Wang, S.L.; Stroe, D.I.; Fernandez, C.; Xiong, L.Y.; Fan, Y.C.; Cao, W. A novel power state evaluation method for the lithium battery packs based on the improved external measurable parameter coupling model. *J. Clean. Prod.* **2020**, *242*, 118506. [CrossRef]

11.   Liu, R.; Chen, J.; Xun, J.; Jiao, K.; Du, Q. Numerical investigation of thermal behaviors in lithium ion battery stack discharge. *Appl. Energy* **2014**, *132*, 288–297. [CrossRef]

12.   Gachot, G.; Grugeon, S.; Eshetu, G.G.; Mathiron, D.; Ribière, P.; Armand, M.; Laruelle, S. Thermal behaviour of the lithiated-graphite/electrolyte interface through GC/MS analysis. *Electrochim. Acta* **2012**, *83*, 402–409. [CrossRef]

13.   Wang, T.; Tseng, K.J.; Zhao, J.; Wei, Z. Thermal investigation of lithium-ion battery module with different cell arrangement structures and forced air-cooling strategies. *Appl. Energy* **2014**, *134*, 229–238. [CrossRef]

14.   Mahamud, R.; Park, C. Reciprocating air flow for Li-ion battery thermal management to improve temperature uniformity. *J. Power Sources* **2011**, *196*, 5685–5696. [CrossRef]

15.   Sheng, L.; Su, L.; Zhang, H.; Li, K.; Fang, Y.; Ye, W.; Fang, Y. Numerical investigation on a lithium ion battery thermal management utilizing a serpentine-channel liquid cooling plate exchanger. *Int. J. Heat Mass Transf.* **2019**, *141*, 658–668. [CrossRef]

16.   Wang, C.; Zhang, G.; Meng, L.; Li, X.; Situ, W.; Lv, Y.; Rao, M. Liquid cooling based on thermal silica plate for battery thermal management system. *Int. J. Energy Res.* **2017**, *41*, 2468–2479. [CrossRef]

17.   Kizilel, R.; Sabbah, R.; Selman, J.R.; Al-hallaj, S. An alternative cooling system to enhance the safety of Li-ion battery packs. *J. Power Sources* **2009**, *194*, 1105–1112. [CrossRef]

18.   Huang, H.; Wang, H.; Gu, J.; Wu, Y. High-dimensional model representation-based global sensitivity analysis and the design of a novel thermal management system for lithium-ion batteries. *Energy Convers. Manag.* **2019**, *190*, 54–72. [CrossRef]

19.   Wang, T.; Tseng, K.J.; Zhao, J. Development of efficient air-cooling strategies for LIB module based on empirical heat source model. *Appl. Therm. Eng.* **2015**, *90*, 521–529. [CrossRef]

20.   Wang, J.; Gan, Y.; Liang, J.; Tan, M.; Li, Y. Sensitivity analysis of factors in fluencing a heat pipe-based thermal management system for a battery module with cylindrical cells. *Appl. Therm. Eng.* **2019**, *151*, 475–485. [CrossRef]

21.   Yang, N.; Zhang, X.; Li, G.; Hua, D. Assessment of the forced air-cooling performance for cylindrical LIB packs: A comparative analysis between aligned and staggered cell arrangements. *Appl. Therm. Eng.* **2015**, *80*, 55–65. [CrossRef]

22.   Ismail, N.H.F.; Toha, S.F.; Azubir, N.A.M.; Ishak, N.H.M.; Hassan, M.K.; Ibrahim, B.S.K. Simplified heat generation model for lithium ion battery used in electric vehicle. *IOP Conf. Ser. Mater. Sci. Eng.* **2013**, *53*, 3–8. [CrossRef]

23.   Chen, S.; Peng, X.; Bao, N.; Garg, A. A comprehensive analysis and optimization process for an integrated liquid cooling plate for a prismatic lithium-ion battery module. *Appl. Therm. Eng.* **2019**, *156*, 324–339. [CrossRef]

24.   Park, H. A design of air flow configuration for cooling lithium ion battery in hybrid electric vehicles. *J. Power Sources* **2013**, *239*, 30–36. [CrossRef]

25.   Du, X.; Qian, Z.; Chen, Z.; Rao, Z. Experimental investigation on mini-channel cooling-based thermal management for Li-ion battery module under different cooling schemes. *Int. J. Energy Res.* **2018**, *42*, 2781–2788. [CrossRef]

26.   Zhao, R.; Gu, J.; Liu, J. An investigation on the significance of reversible heat to the thermal behavior of lithium ion battery through simulations. *J. Power Sources* **2014**, *266*, 422–432. [CrossRef]

*Article*

# Connecting Parking Facilities to the Electric Grid: A Vehicle-to-Grid Feasibility Study in a Railway Station's Car Park

Ruben Garruto [1], Michela Longo [1,*], Wahiba Yaïci [2,*] and Federica Foiadelli [1]

[1] Department of Energy, Politecnico di Milano, via La Masa, 34–20156 Milan, Italy;
ruben.garruto@mail.polimi.it (R.G.); federica.foiadelli@polimi.it (F.F.)

[2] CanmetENERGY Research Centre, Natural Resources Canada, 1 Haanel Drive, Ottawa, ON K1A 1M1, Canada

* Correspondence: michela.longo@polimi.it (M.L.); wahiba.yaici@canada.ca (W.Y.);
Tel.: +39-02-2399-3759 (M.L.); +1-613-996-3734 (W.Y.)

Received: 24 March 2020; Accepted: 9 June 2020; Published: 15 June 2020

**Abstract:** This study evaluates the impact of energy on the distribution network at the point of connection of an electric plant of a railway car parking facility in which charging points for electric vehicles (EVs) were installed. The objective is to identify a possible load curve of the simulated car park and, based on the principle of vehicle-to-grid (V2G) technology, to develop an appropriate algorithm. Such an algorithm explores the possibility of a two-way energy flow between the connected vehicles and the electricity grid, and performs a peak shaving of the load curve of the plant under examination in order to avoid absorption peaks, which are usually difficult to manage when using the distribution system operator (DSO). The work also presents the coupling with a photovoltaic system designed specifically for the car park. The study results are presented after a summary of the current state of development of electric mobility, describing the various types of EVs, the charging infrastructure, and the possible applications in smart grids (SGs).

**Keywords:** electric vehicles (EVs); photovoltaic (PV) systems; vehicle-to-grid (V2G); smart grids (SGs); peak shaving

---

## 1. Introduction

Electric mobility represents a necessary transition to extricate from a mode of transport that depends on the use of harmful fossil fuels, of which emissions into the environment make living in urban centers increasingly hazardous. Electric mobility, being one of the branches of "smart mobility", is therefore a part of a pillar of "smart cities", which are a congregation of urban planning strategies and plans aimed at ensuring sustainable economic development and a high quality of life through a skillful management of available resources (food, mining, energy, services, etc.) [1,2].

However, it would be reductive to limit "smart nobility" by merely switching from a type of transport with an internal combustion engine to one with an electric motor. The reality is that "smart mobility" is a much broader concept involving different stakeholders (public administrations, private/public or mixed companies, end users, etc.) and technologies/services (car sharing, autonomous driving, trip planning apps, etc.). A comprehensive review is outside the scope of this study, which deals instead with the concept of vehicle-to-grid (V2G), as previously mentioned. Nowadays, in reality, the concept of the electric vehicle (EVs) is still of the "passive" type, of the grid-to-vehicle type (G2V), conceived to mere electric load. However, this paradigm could be destined to change in the immediate future. The idea of V2G fits into the more general concept of the "smart grid" (SG).

The current electrical system originated at the start of the 20th century and was conceived as an exclusively unidirectional system [3,4].

Going back in time, debates about the desirability of migrating to electric mobility commenced at the end of the 19th century. Even before problems associated with fossil fuel use in transportation began to manifest, a jury announced its introduction in 1898, following a contest in which several prototypes of cars (both electric and with an internal combustion engine) were presented.

Indeed, the early 1900s witnessed the evolution of this unique electricity-powered mode of transportation [5–7]. Electric cars seemed preferable to petrol cars, which were noisy, had annoying vibrations and smoky exhausts with attendant risks of fire and explosion. Furthermore, the manual gearbox and cranking made them difficult to use. Although the first electric cars could only cover minimal distances of 80–90 km, this did not constitute a problem since the movements at the time were confined to domestic destinations, at least in the initial phase. However, that mild status seemed destined for change following the massive production of the famous T model of Tesla, a reduction in oil price, and improvements in road networks, which enabled greater distances to be covered. An alternative method of transport with an internal combustion engine became a necessity [8–10].

All these factors contributed to a change in direction towards fossil fuel mobility, which became predominant throughout the 20th century up to the present day. The oil crisis of the 1970s and the resultant economic crisis, however, highlighted the vulnerability of an economic system that depended heavily on fossil fuels, which, in addition to their obvious limitation of heavily polluting the environment, tended to be found predominantly in regions that were plagued with profound political instability. The crisis prompted world governments to radically modify their assets with the aim of reducing the West's dependency on Middle Eastern oil sources. This would be accomplished by reduced consumption, the reorganization of industrial production, and a search for alternative energy sources. The crisis clearly also involved the transport sector, and the migration to an electric transport mode—which, until then, had been confined to specific applications—began to attract considerable interest again [11,12].

As stated in the introduction to this paper, it is still premature to talk about an "electrical revolution". However, the subject of environmental protection is increasingly topical, and coupled with the reality that fossil fuels will likely become a rare resource, thus, the switch to an alternative in the form of electrical mobility has become a topic of considerable debate over the last ten years.

As stated in [1], regarding the current Europe panorama, the transportation sector occupies the fourth place as one of the largest emitters of greenhouse gases with a share of 14%. Increased use of electric mobility would certainly reduce this proportion and mitigate the critical problem of pollution in urban centers. Despite the initial problems and criticalities linked to the migration from thermal to electric mobility, a multigovernmental initiative known as the Electric Vehicles Initiative (EVI) was inaugurated in 2009 to accelerate and control the deployment of EVs worldwide [13]. Another recent initiative proposed by the EVI is the EV30@30 [2] Campaign, which was launched at the Eighth Clean Energy Ministerial in 2017 in Beijing (China). The aspirational goal of this campaign is to achieve 30% sales share for EVs by 2030. The main actions to attain this goal include the following: (i) provide support to the governments in need of policy and technical assistance; (ii) promote programs (such as the Global EV Pilot City Programme) to facilitate the exchange of experiences in the EV field, and propose the best practices for the promotion of EVs in cities; (iii) encourage public and private sector commitments for EV uptake in company and supplier fleets; (iv) support the deployment of EV chargers and track progress. In 2017, the global stock of electric passenger cars reached 3.1 million, which was a 57% increase on the number for 2016 and approximately two-thirds of the world's electric car fleet are battery-operated electric vehicles (BEVs).

In addition to the 3.1 million passenger electric cars, there were nearly 250,000 electric light commercial vehicles (LCVs) on the road in 2017. The largest electric LCV fleet is in China (170,000 vehicles), followed by France (33,000 vehicles) and Germany (11,000 vehicles). Electric LCVs are often part of a company or government fleet. For instance, the DHL Group, which is a major logistics company, operates with the largest EV fleet in Germany with 16,000 electric vans, bicycles and tricycles. The company has also undertaken in-house development and manufacturing of its own electric vans, tricycles, and bicycles as part of this vision. Following its success in this venture, the company is now selling its EVs to third parties (mainly municipalities and other businesses).

The year 2017 recorded over one million sales of EVs worldwide, which was an increase of 50% in comparison to the previous year. Despite this, it is acknowledged that this is still a modest figure to refer to as an "electric mobility revolution". The two leading nations in this sector are Norway and China, with different merits. China has the largest car market and nearly 580,000 electric cars were sold there in 2017. On the other hand, Norway can boast a greater market penetration of EVs, as 39% of vehicles sold in 2017 were electric.

In addition to e-mobility, it is necessary to give a brief overview of smart-mobility, which represents all the new possibilities of using the car and its interaction with the network [14]. On average, electric cars, due to their charging and autonomy characteristics, are used for their own movement and transport activities for very short periods, that is, they are in movement for just 5% of the time, while they remain stationary and therefore unused for the remaining 95% of time [15]. Given their flexible load when electric storage available, as well as the development of communication systems, it would be irrational to view them as just a regular load [16,17]. Their main function will remain in the transportation sector, but with a high potential in providing services, both to the grid and to the market [18].

The V2G mode of operation can be applied, exchanging energy in the system during critical conditions, and since electric vehicles are dispersed by nature, this increases their impact on grid regulation, facilitating the work of the network operator [19]. This application for EVs can reduce the overall costs of purchasing and maintaining storage units in the future, and reduce the need for grid upgrades, which would result in an economic benefit for the Distribution System Operator (DSO). Consumers providing this service through their vehicles, of course, should be properly compensated, with the possible creation of new markets [20,21]. However, the management of this resource is extremely complex, with some elements requiring strong regulation. Furthermore, charging and discharging the battery for V2G increases the cycles carried out, causing greater degradation to the battery. The owner must also be paid for this reason, and must be informed about the amount of the remaining charge to be sure of being able to use the vehicle after having transferred energy to the network. The fee will depend on the market price and the time in which the vehicle remains available for the service [22,23].

Finally, vehicles must communicate with the electrical system via a special connection inside the plug. From this analysis was born the idea of a full time use of electric vehicles, using them not only for traditional activities but also as energy accumulators. In the energy management process, the role played by storage systems is certainly complementary to renewable generation and contributes in giving stability to the network, both as a quick response to the required needs and as a reduction in power fluctuations related to renewable energy generation produced by other sources, such as photovoltaics and wind [24,25].

The latter, in fact, being dependent on climatic factors and therefore not programmable, cannot provide optimal and immediate solutions to sudden requests from the network, so an efficient storage system is essential to manage the differences between production and use. By taking advantage of the charged batteries of electric cars connected to the network, it is possible to offer, in a very short period of time, an adequate response to a sudden peak in demand. If on one hand the diffusion of electric vehicles could contribute to the stability of the network by regulating the frequency and reducing the power fluctuations linked to renewable generation, on the other hand, the uncontrolled growth of electric cars, not adequately supported by an expansion of the distribution network, could produce an opposite effect [26]. According to the simulations performed, the current network infrastructure has the potential to support a medium–low EV penetration (below 50%). In particular, considering the trend of the load in the tertiary sector, the effects of electric vehicles are not significant since the requests of recharge are concentrated in the morning and are not summed up to the typical afternoon peak.

In the residential area, however, the effect of electric vehicles affects the network more heavily, since the recharges are concentrated mainly in the evening, thus producing greater peaks. It is possible to make the vehicles interact with the grid at different levels with several options, such as V2H (vehicle-to-home), V2B (vehicle-to-building) and V2G. In this way, the bidirectionality of the battery is exploited to allow an energy flow from the vehicle to the network. Vehicle batteries store a lot of energy, on average about 25 kWh (but those of the latest generation reach up to 90 kWh), which correspond to about two and a half days of energy required by normal users. It is therefore a big amount of energy that can be used. Taking advantage of this capacity is not easy, and it is convenient only when large quantities of energy are involved or when many vehicles are in the same place, located together as the only source such as in parking lots, near charging stations and in residential areas during the night. V2G can be used to provide power supply at peak demand for an intervention time varying between four and six hours every day, or for ancillary services in order to occasionally supply power to support the system in case of frequency variations due to loss of generation or problems on transmission lines. In this case, these are single interventions with a limited average duration of ten minutes. Finally, regulation can also be performed in order to exchange active and reactive power to guarantee the voltage regulation of the system. The amount of operations that can be performed is very high, but they can only be performed for short time intervals of a few minutes at a time.

V2G energy transfer has been extensively investigated in various nations to mitigate fluctuating demands and supply availability changes. For instance, Perujo and Ciuffo [27] assessed the introduction of electric vehicles in the private fleet with regard to potential impact on the electricity supply and on the environment for a case study for Milan, Italy. Ekman [28] investigated the interaction between large EV fleets and high wind penetration in Denmark. Hartman et al. [29] assessed the influence of various exploitation scenarios of EVs on the German grid in 2030. Drude et al. [30] examined photovoltaics and V2G approaches for peak demand reduction in Brazilian urban regions in a smart grid environment. Similar studies were performed for other areas in the world [31–34].

Moreover, as disclosed in a literature review, several researchers have dealt with V2G and renewable energy source integration [35–39]; smart grid operation considering large-scale integration of EVs enabling V2G systems [40,41]; peak shaving and valley filling of power consumption using an EV parking lot [42,43]; optimization of integration of EVs in SG and EV charging stations [44–49].

As per the authors' understanding, it appears that no V2G feasibility study or alternative form of work has been accomplished on connecting the car park of Ferrara railway station in the region of Emilia-Romagna, northern Italy, to the SG.

Therefore, considering the V2G concerns and challenges in the context of the SG described above, this study aims to evaluate the impact of energy on the distribution network at the point of connection of an electric plant of a railway car park in which some charging points for EVs were installed. Case studies are presented after an overview of peak shaving and load levelling, descriptions of the various types of EVs, the charging infrastructure and the possible applications in SGs. The objective is to identify a possible load curve of the simulated car park and, based on the principle of V2G technology, an appropriate algorithm will be developed. This algorithm will explore the possibility of a two-way energy flow between the connected vehicles and the electricity grid, and perform a peak shaving of the load curve of the plant under examination in order to avoid absorption peaks which are normally difficult to manage by the distribution system operator (DSO). The work also presents the coupling with a photovoltaic system designed specifically for the car park.

## 2. Overview of Peak Shaving and Load Levelling

The management of the electricity grid is very complex, since the power consumption by the users is characterized by very marked fluctuations [14,15]. Peak production by Renewable Energy Sources (RES) such as solar panels does not correspond with the peak loads of the system, so storing that energy in the electrical vehicles would be suitable for guaranteeing system adequacy, reducing the need of investments in the generation sector. In addition, the stability of the system might be endangered due to a high percentage of uncontrollable loads. The use of an electrical energy storage system according to the load levelling logic consists in the storage of energy during the low load hours of the electrical system and its subsequent supply during periods of high demand from users, as simulated in the QDSL battery model [16,17]. This operating criterion entails, above all, technical benefits, in particular by optimizing the operation of the thermal plants and the management of the electricity network, in addition to the peak shaving operation, which is useful for levelling the peaks of maximum electricity demand. Furthermore, the intervention of a storage system can satisfy the peak demand while maintaining a more homogeneous generation profile. Thermal plants, in particular those that exploit low-cost primary sources, are not designed to operate at a partial load with good efficiencies. By accumulating the surplus of electricity that they generate during periods of low demand, it is possible to allow these systems to always operate near nominal load and efficiency. The stored energy can be subsequently supplied during the hours of maximum demand by reducing the load of expensive peaking systems powered by natural gas. Higher penetration of EVs means higher electrical energy requirements, as well as bigger stress on the electrical grid. The storage systems can intervene, as previously described, to level the load curve and manage the periods of maximum demand by postponing investments on new network infrastructures and on increasing installed power (Figure 1). This operating strategy also favors the practice of energy price arbitrage. It takes advantage of the price difference of electricity that occurs between two times of the day or week to make an economic profit for the storage system. Electricity is purchased and stored when it costs little due to the low demand from users, and then it is resold during the hours when prices are higher. From an economic point of view, it is possible to affirm that energy storage systems favor the reduction and stabilization of prices in the electricity market because they release electricity production from speculation and the volatility of prices linked to fossil fuels.

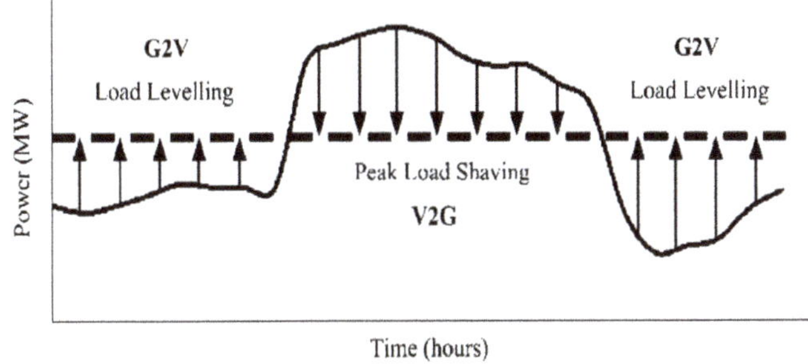


**Figure 1.** Peak shaving and load levelling example.

## 3. Case Study

The location of this study is the parking lot situated adjacent to the train station. The area is marked by an intensive one-modal exchange for the presence of the railway station, bus stops and terminals, bike parking and taxi stops, as shown in Figure 2. It is expected that a new parking lot will soon be located adjacent to the railway station and will consist of 115 parking spaces, 5 of which will be equipped for charging EVs. The total area occupied will be 4300 m$^2$.

In order to estimate the car park's energy consumption and then derive the relative load curve, it is necessary to determine the electrical loads belonging to the parking lot. After that, a hypothetical load curve is developed in for each utility in the most unfavorable conditions from a grid absorption point of view. The fusion of the various load curves is represented by the effective parking absorption curve which will ultimately be the reference energy consumption used during the successive analysis.

**Figure 2.** Top view of the parking lot.

### 3.1. Lighting Systems

Lighting for the parking lot is optimized by the provision of 18 lampposts, which run along two main panels across the entire length of the parking lot. Each lamp's strength is 74 W, which aggregates to a total of 1.3 kW. For the most unfavorable case, December 21st was chosen because it is the shortest day of the year. It was hypothesized as a continuous operation that goes from 4:30 p.m. at 8:00 a.m. The resulting power curve is presented in Figure 3.

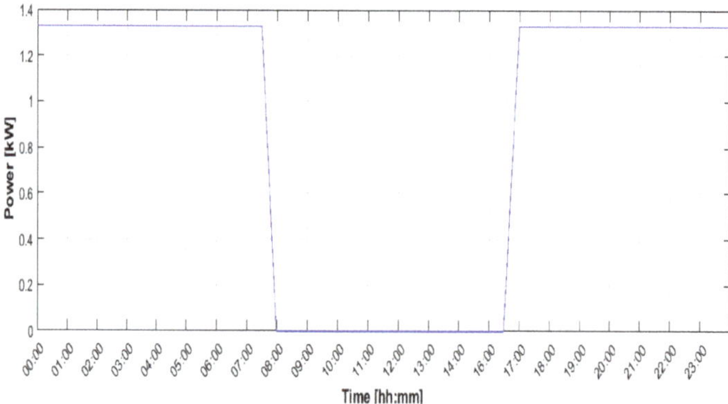

**Figure 3.** Load curve of the lighting system.

### 3.2. Entry/Exit System

The parking access points consist of an input gate and an output gate with a rated output of 650 W each. In order to realize the load curve, it was assumed that the entrance gates will be most frequently used between 6:30 a.m. and 9:00 a.m. Similarly, the second peak time was assumed to be between 4:30 p.m. and 8:30 p.m. This hypothesis is quite plausible considering that railway parking lots are mainly patronized by commuters. Thus, it can be assumed that there would be greater movement in and out of the car park's access and exit gates within the time bands where people go to their workplaces and when they return to go home. The relative load curve is represented in Figure 4.

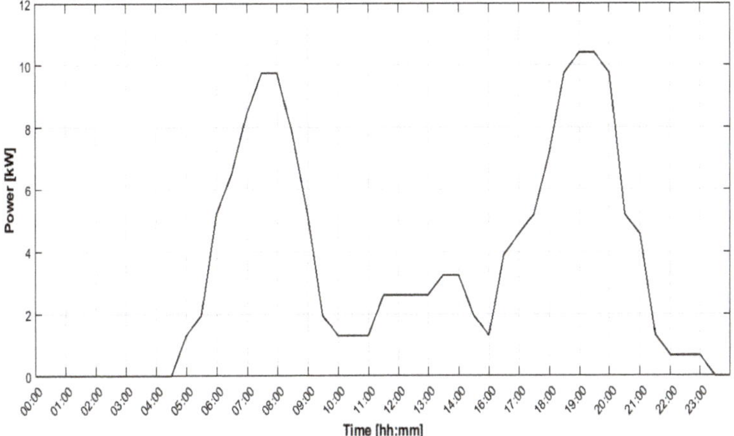

**Figure 4.** Load curve of the entry/exit system.

### 3.3. Video Surveillance System

The parking lot is equipped with a 24-h video surveillance system. The total number of cameras is nine; each camera absorbs a rated power of 5 W for a total absorption of 45 W. This is a very modest load when compared to the previous loads analyzed. The load curve shows a consistent, trend as indicated in Figure 5.

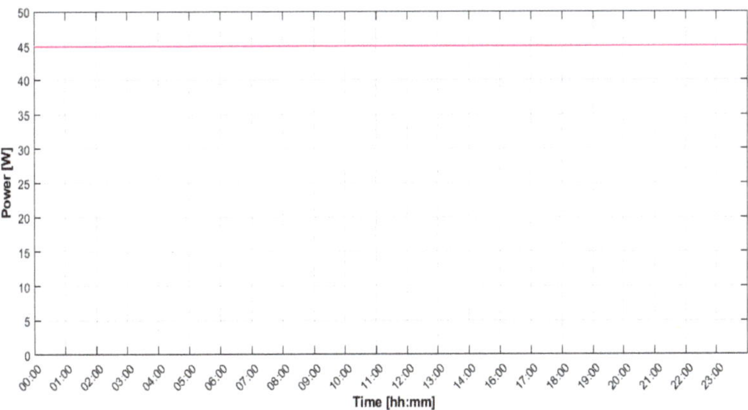

**Figure 5.** Load curve of the video surveillance system.

### 3.4. Technical Room

The parking area has a technical room available for use by parking operators. The main load that can be hypothesized is the air conditioning system, which is attested on a power of about 800 W. In order to factor in accessory loads, a constant absorption of 1000 W is assumed over the entire period of service of the car park. Figure 6 indicates the expected trend.

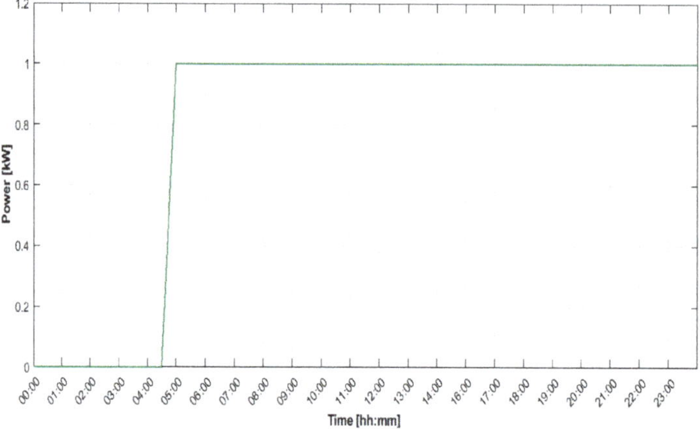

**Figure 6.** Load curve of the technical room.

## 3.5. Electric Vehicle Supply Equipment

Another load to introduce is the one constituted by the EV supply equipment. The car park in question allows charging for five vehicles at a time. An uncontrolled connection of gives electric vehicles, albeit modest in number, goes to engage the distribution network. Figure 7 assumes the worst case, that is, the simultaneous presence of five EVs that require a complete recharge (substantially, from 0% to 100%) in the load peak.

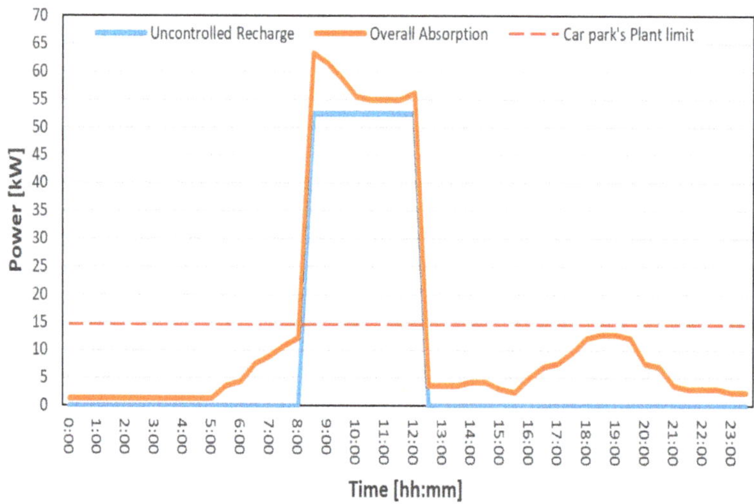


**Figure 7.** Example of uncontrolled charging of the five EVs implemented in the car park.

## 3.6. Electric Vehicle

The reference car used as EV is the Nissan Leaf 2018. It was selected because that model is one of the few electric cars certified for use in V2G applications [50]. Table 1 shows some of the car's technical characteristics. The Nissan Leaf features two charging sockets—Type 2 and CHAdeMO. The on-board charger has a rated power of 6.6 kW, while the CHAdeMO socket supports a rated output of up to 50 kW.

**Table 1.** Electrical specification data for Nissan Leaf 2018 electric vehicle (EV) [51].

| Parameter | Value | |
|---|---|---|
| Battery capacity | 40 | [kWh] |
| Charge power/Type 2 | 6.6 | [kW], 1 or 3-phase |
| Fastcharge port/CHAdeMO type | 50 | [kW], DC |
| Input voltage | 360 | [V] |
| Vehicle consumption [1] | 236 | [Wh/km] |

[1] "worst case" based on −10 °C and use of heating along a highway.

Concerning the effective charging power, it is important to keep in mind that the discriminant is the minimum value of the power between the supply equipment and the charger of the vehicle itself according to the following relation (Equation (1)):

$$P_{charge} = min(P_{EVSE}, P_{EV})$$

(1)

with:

$P_{charge}$: rated power of recharge;

$P_{EVSE}$: rated power of the electrical vehicle supply equipment;

$P_{EV}$: rated input power of the electrical vehicle charger.

The maximum power absorbed by the network that will be considered is that of the electric vehicle supply equipment (EVSE), which is equal to 10.5 kW. Of the 10.5 kW, 10 kW is actually used to recharge the vehicle, while the remainder represents the losses of the charging system. Figure 8 indicates the trend of the power absorbed by the grid at three different power levels, based on standard value.

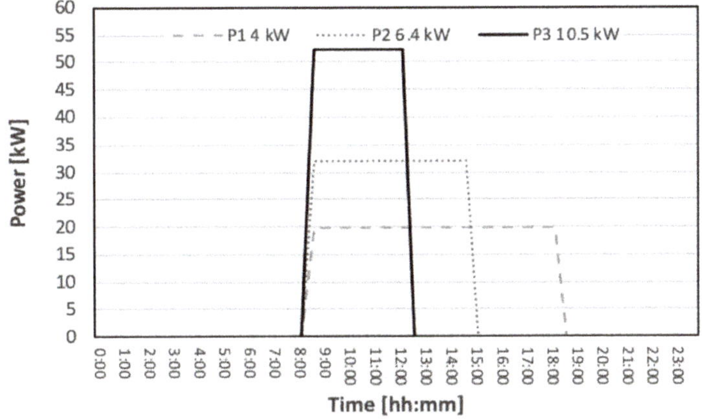


**Figure 8.** Charging load at different power levels considering five Nissan Leaf EVs.

## 4. Analysis of the Energy Absorption of the Car Park

As seen so far, the car park under examination is equipped with a series of electrical loads, which, overall, indicates power absorption similar to that shown in Figure 9. The red dotted line represents the maximum power limit that can be absorbed by the car park's electrical system, which stands at around 14 kW. Figure 9 shows the basic load of the car park. Now, assuming that power absorbed by electric cars is added, it will derive a load curve that is similar to that in Figure 7, which assumed the simultaneous recharge of five EVs during the first peak. However, this assumption presents a rather burdensome situation, because it requires a considerable amount of energy to be absorbed from the grid over an extensive period. This is a classic example of an uncontrolled charging without the implementation of any smart charging feature.

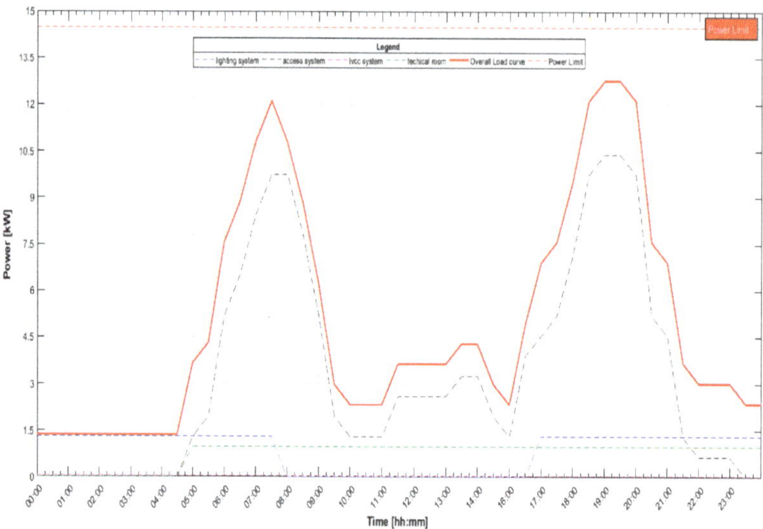

**Figure 9.** Overall load curve of the car park.

The relative energy consumption is shown in Figure 10. Appreciably, greater energy absorption occurs between 7.00 a.m. and 10.00 a.m. and between 5.00 p.m. and 9.00 p.m., and as a result of which, two peaks of power are derived. The daily total energy absorbed is approximately 114 kWh.

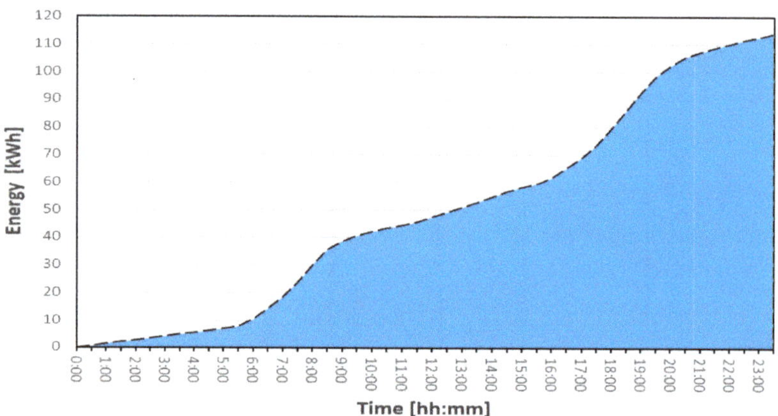

**Figure 10.** Overall daily energy absorption of the car park (without Electric Vehicle Supply Equipment—EVSE).

Table 2 shows a summary of the overall energy absorption at the car park when deploying a full charge of five Nissan Leaf cars for every day of the year.

**Table 2.** Overall energy absorption of the car park.

| Parameter | Basic Consumption | Consumption with EVs |
|---|---|---|
| Daily energy absorption [kWh] | 118.2 | 328.2 |
| Monthly energy absorption [MWh] | 3.55 | 9.85 |
| Annual energy absorption [MWh] | 43.1 | 119.8 |

## 5. Design of the Photovoltaic System

One of the major criticisms also giving rise to skepticism about electric mobility is that the energy to recharge this fleet of vehicles must be sourced from somewhere. This point is not trivial should fossil fuels constitute and remain the source of this much-needed energy for the recharging of vehicles. It means that the thorny problem of environmental pollution remains unresolved, and the problem is transferred to the supply chain. In fact, if one considers the tank-to-wheel transformation chain, the $CO_2$ emissions into the atmosphere of both the EV and the Internal Combustion Engine (ICE) become equivalent.

However, the problem does not arise (or, is mitigated) if instead, the energy used came from a renewable source. Therefore, concurrently with the development and widespread use of the EVs, it is important to launch initiatives that aim at a greater diffusion of renewable energy sources [52,53]. Following from this, a photovoltaic power plant will be designed in this section in order to provide the car park's load demand. As already described, the reference car park is the parking lot adjacent to the railway station. In this car park, the installation of three EVSEs with V2G technology was planned. Geographically, the site of interest has an azimuth of 30° towards the west, as shown in Figure 11. The optimal tilt angle of the panels was instead set at 30°.

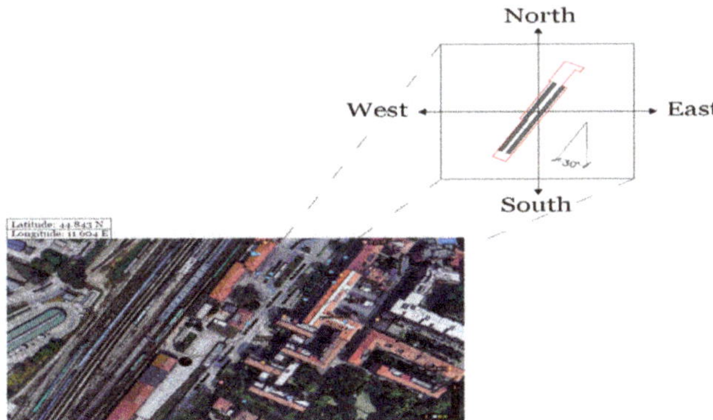


**Figure 11.** Bird-eye view for the site of interest—the car park.

The average daily solar radiation of this site, characteristic for each month, was obtained from [52]. From these data, the graph depicted in Figure 12 was derived.

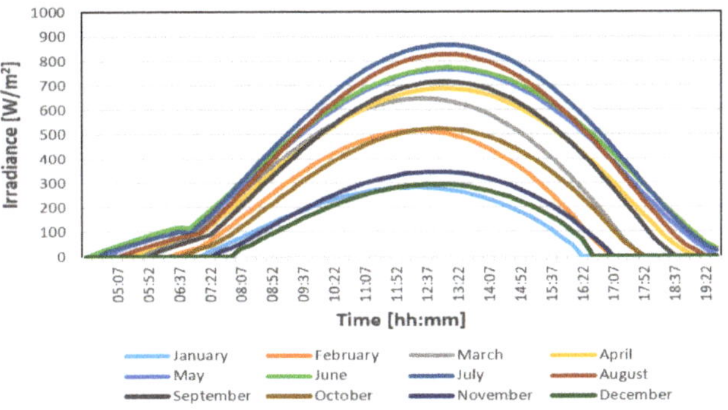


**Figure 12.** Daily irradiance in Ferrara for each month.

From the solar irradiance dataset, exploiting Equation (2) [53], the associated produced power can be computed as

$$P_{out,panel}(G_t) = \begin{cases} P_n \cdot \left( \dfrac{G_t^2}{G_{std} \cdot R_c} \right) \text{ for } 0 < G_t < R_c \\ P_n \cdot \left( \dfrac{G_t}{G_{std}} \right) \text{ for } G_t > R_c \end{cases} \qquad (2)$$

where:

$P_{pv}(G_t)$: the output power from a single panel as a function of the solar irradiance $G_t$;

$G_t$: the forecasted solar irradiance measured in W/m$^2$ at a certain time t in a day;

$P_n$: the nominal output power of the photovoltaic panel chosen;

$G_{std}$: solar radiation in the standard environment set as 1000 W/m$^2$;

$R_c$: a certain radiation point set as 150 W/m$^2$.

Considering that the data have been sampled at intervals of a quarter of an hour from one another, the energy can be easily obtained through Equation (3):

$$E_t = P_{out,panel}(G_t) \cdot \Delta t \qquad (3)$$

*5.1. Photovoltaic (PV) Panel Technology*

The PV panel selected for this study case is the panel NeON® 2 BiFacial. The peculiarity of this panel is its double-sided structure which is able to exploit both direct light and reflected light. It is possible to appreciate the differences between a conventional photovoltaic module and one considered with double-sided technology. The bifacial cell is designed in a symmetrical structure in order to gain additional sunlight absorption from the rear.

The panel therefore uses the Albedo effect [48]. The Albedo is an index derived from the ratio between the light incident on a surface and the corresponding reflected light. Consequently, the Albedo index can vary between 100% for perfectly reflective surfaces and 0% for perfectly absorbent surfaces.

In a context such as the one under examination, it is assumed that the surface is mostly made up of asphalt, which corresponds to an Albedo between 10% and 20%. A power gain of 5% with respect to the nominal power is therefore assumed.

*5.2. PV Support Structure*

The project involves the installation of specific support structures for the PV modules to allow an optimal exploitation of the available surface. The proposed structure is illustrated in Figure 13. It is a modular solution that simultaneously occupies two parking spaces and can also hold up to 9 PV panels at a time. The steel structure guarantees the best balance between lightness and strength of the structure. Evidently, the presence of the support rails, due to their shadowing, implies an inversely proportional impact on the power gain of the double-sided technology. For the present work, a gain in power of 5% will be cautiously considered.

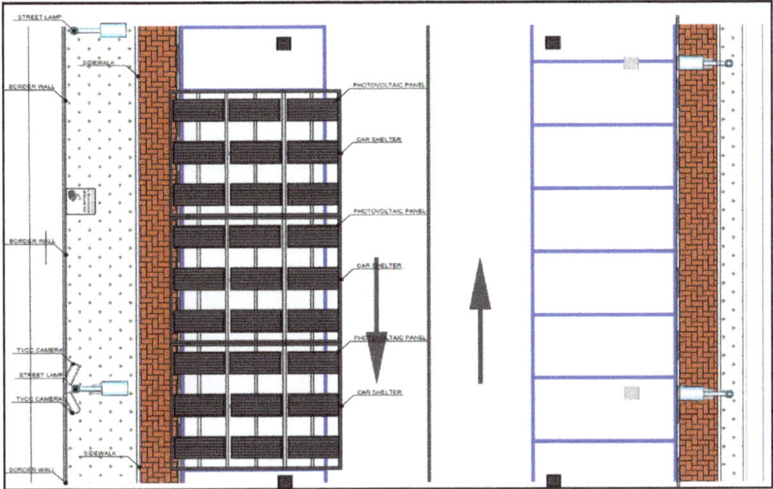

**Figure 13.** Concept of installation (top view).

## 5.3. PV Matching Panels/Inverters

The car park has 114 parking spaces. The maximum number of shelters will therefore be 57 for a total of 513 PV panels to be installed. The maximum peak power of the generation plant will therefore be equal to 200 kWp. The plant was divided into three inverters of equal size, each with a rated power of 70 kW. 171 PV panels will be connected to each inverter, which are divided into 10 parallel strings. 9 strings consist of 17 panels in series and the 10th consists of 18 panels. Table 3 presents a summary table of these data. It proceeds with the verification of correct matching string/inverter, considering the limit temperatures of +70 °C and −10 °C. The conditions are summarized in Table 3 as well.

**Table 3.** String/inverter matching conditions.

| Case | String | | Inverter | Condition |
|---|---|---|---|---|
| 1 | $V_{MPP,min,STR}$ | $>$ | $V_{activation}$ | The minimum output voltage of the string (the one corresponding to +70 °C) must be greater than the activation voltage of the inverter. |
| 2 | $V_{MPP,MAX,STR}$ | $<$ | $V_{MPP,MAX,INV}$ | The maximum output voltage of the string (the one corresponding to −10 °C) must be lower than the maximum Maximum Power Point (MPP) input voltage tolerable by the inverter. |
| 3 | $V_{OC,STR}$ | $<$ | $V_{MAX,INV}$ | The maximum open circuit voltage of the string (the one corresponding to −10 °C) must be lower than the maximum input voltage tolerable by the inverter. |
| 4 | $\sum_{k=1}^{N} I_{MPP,STR-k}$ | $<$ | $I_{MPP,INV}$ | The sum of all the N-string currents must be lower than the maximum input current tolerable by the inverter. |
| 5 | $\sum_{j=1}^{M} P_j$ | $<$ | $P_{n,INV}$ | The total power of the M panels connected to the inverter must be less than its nominal power |

In particular:

$$V_{MPP,min,STR} = N_{panels} \cdot [V_{MPP}(25\,°C) + (70\,°C - 25\,°C) \cdot \Delta V_T] \tag{4}$$

$$V_{MPP,MAX,STR} = N_{panels} \cdot [V_{MPP}(25\,°C) + (-10\,°C - 25\,°C) \cdot \Delta V_T] \tag{5}$$

$$V_{OC,STR} = N_{panels} \cdot [V_{OC}(25\,°C) + (-10\,°C - 25\,°C) \cdot \Delta V_T] \tag{6}$$

Therefore, after all the calculations are completed, the matching conditions are verified and are as reported in Table 4.

**Table 4.** String/inverter matching conditions and calculations.

| String | | Inverter |
|---|---|---|
| $V_{MPP,STR} = 18 \cdot 41.4 = 745$ V @25°C | | |
| $V_{OC,STR} = 18 \cdot 49.2 = 885.6$ V @25°C | | |
| $V_{MPP,min,STR} = 637.6$ V @70°C | > | $V_{activation} = 350$ V |
| $V_{MPP,MAX,STR} = 828.9$ V @−10°C | < | $V_{MPP,MAX,INV} = 960$ V |
| ✔   $V_{OC,STR} = 969.3$ V @−10°C | < | $V_{MAX,INV} = 1000$ V |
| ✔   $\sum_{k=1}^{10} I_{MPP,STR-k} = 94.3$ A | < | $I_{MPP,INV} = 120$ A |
| ✔   $\sum_{j=1}^{171} P_{pk,j} = 66.7$ kW | < | $P_{n,\,INV} = 70$ kW |

### 5.4. Producibility of the Designed PV Plant

Once the system is dimensioned, it is possible to evaluate the producibility, that is, the energy that this plant is able to generate in the calendar year. We have seen that the power generated by a single panel is given by Equation (2), which does not take into account any loss factors. Thus, in order to evaluate the actual power generated, Equation (7) must be considered:

$$P_{out} = \eta \cdot N_{panels} \cdot P_{out,panel}(G_t)$$
$$\eta = \eta_{el} \cdot \eta_{Mismatching} \tag{7}$$

where

$\eta_{el}$: a factor including all electrical losses (cables, inverters, etc.);

$\eta_{Mismatching}$: a corrective factor affecting the output power of the panels due to various causes, including the difference in thermal gradient of the modules, different shading of the modules for passing clouds, accumulation of dirt, intrinsic differences of the modules, etc.

A correction factor $\eta = 0.85$ was assumed for this project. Figure 14 presents an estimate of the power supplied and the relative energy that can be produced by the PV plant.

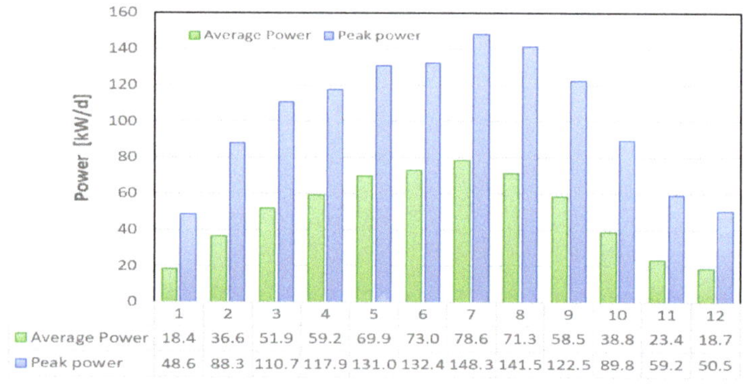


(**a**)

**Figure 14.** *Cont.*

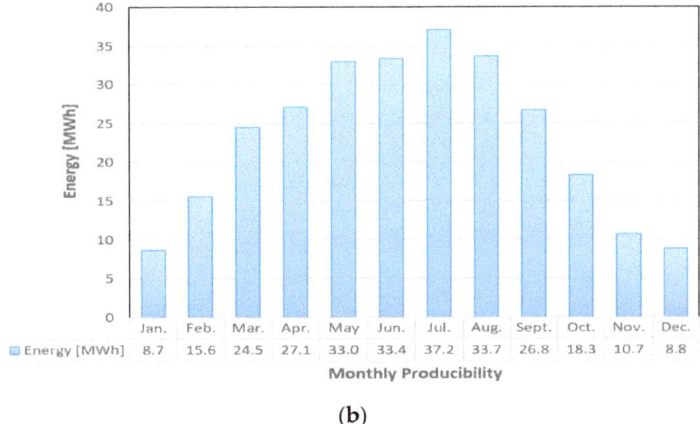



**(b)**

**Figure 14.** (**a**) Daily output power of the photovoltaic (PV) plant month by month and (**b**) overall monthly producibility of the PV plant.

## 6. Smart Charging of the Electric Vehicles

A possible algorithm for optimizing the charging process of EVs is presented. As discussed earlier in this study, the implementation of the smart charge for EVs is mandatory in order to avoid situations of dangerous overloads in the distribution network. However, the management of energy flows is only possible in a mature context that has a reliable and stable smart grid. In the following paragraph, two separate optimization algorithms, which exploit two different principles, are proposed:

"Green" algorithm: the first algorithm proposed is based on a very simple principle. Given the presence of a photovoltaic system on the site of interest, the charging of the EVs is entrusted only to this last resource. That way, their recharge does not affect the distribution network. It is easy to see that if the generation system does not work (at night, or on a cloudy day), this algorithm will no longer be valid. In this case, we must consider a different principle such as that which will be explained later.

"Peak shaving" algorithm: this algorithm is applicable regardless of the presence of a renewable resource. Based on the energy absorption in the network and the presence of the parked EVs, the controller will decide whether to extract energy from the vehicles in order to reduce the peak rate of energy absorption, or whether to recharge them. The ultimate goal is therefore to level the power peaks and obtain a load curve that is as smooth as possible.

### 6.1. Green Algorithm

This algorithm allows one to take advantage of the renewable resource in order to recharge the EVs. Its basic principle is illustrated in the flowchart of Figure 15 and it is structured in a manner that allows recharging only when energy comes from the photovoltaic system, otherwise the vehicles are not recharged. In this way, the recharging phase minimizes the impact on the distribution network. Let us therefore see a practical application of the "green" algorithm to the case in question. In the previous paragraphs, both the energy absorption of the car park under analysis and the photovoltaic production of the designed power plant were analyzed.

Figures 16 and 17 present a comparative analysis of both. As a reference month, January was chosen because it has the lowest amount of energy produced and thus represents the most critical case. The presence of a fleet of five vehicles was also assumed. A 25% charge was set for each vehicle. In Figure 18, the orange boxes indicate the load curve of the parking lot, while the green boxes indicate the power generated by the photovoltaic system. At a glance, even considering January's low producibility, photovoltaic production appears to effectively compensate for the parking energy demand, thus leaving a good margin of power for recharging the EVs.

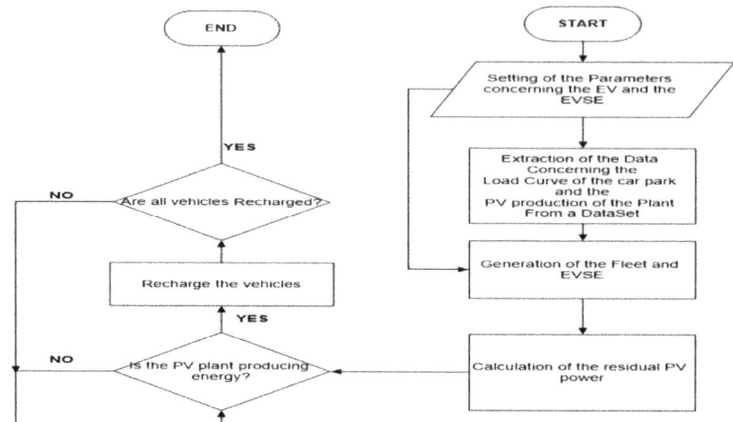


**Figure 15.** Flowchart of the "green" algorithm.

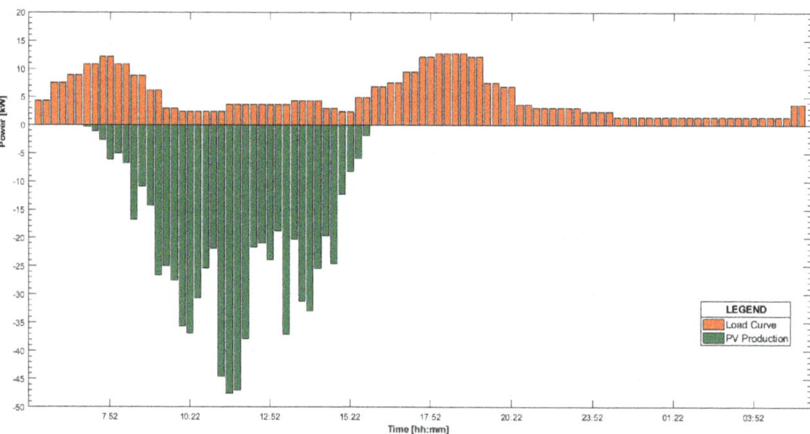


**Figure 16.** Comparison of load curve of the car park and PV production distributions.

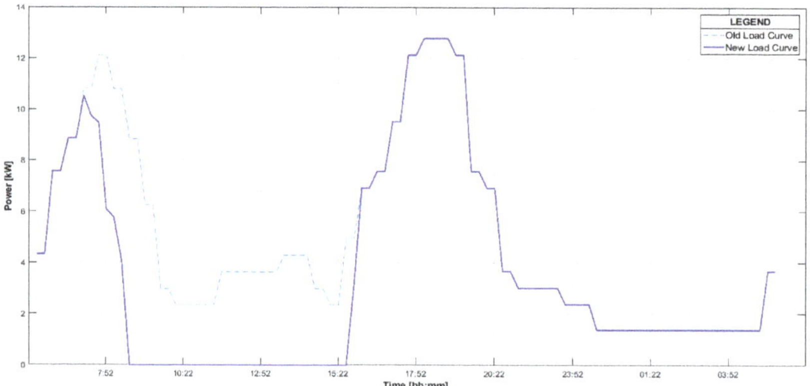


**Figure 17.** Comparison between the old load curve and the new one.

Figure 18 presents the load curve before and after the energy introduced by the PV plant has been used.

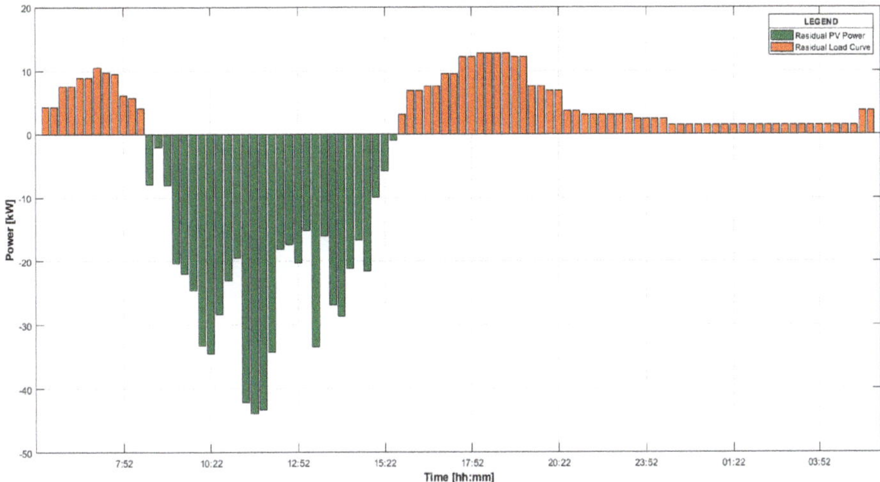

**Figure 18.** Net power production and power absorption distributions.

Due to the residual energy produced, it is possible to recharge the fleet of vehicles in approximately 5 h, as Figure 19 indicates. The red curve shows the power used to completely recharge the fleet composed of five vehicles. This power is not absorbed by the network, but is entirely produced on site.

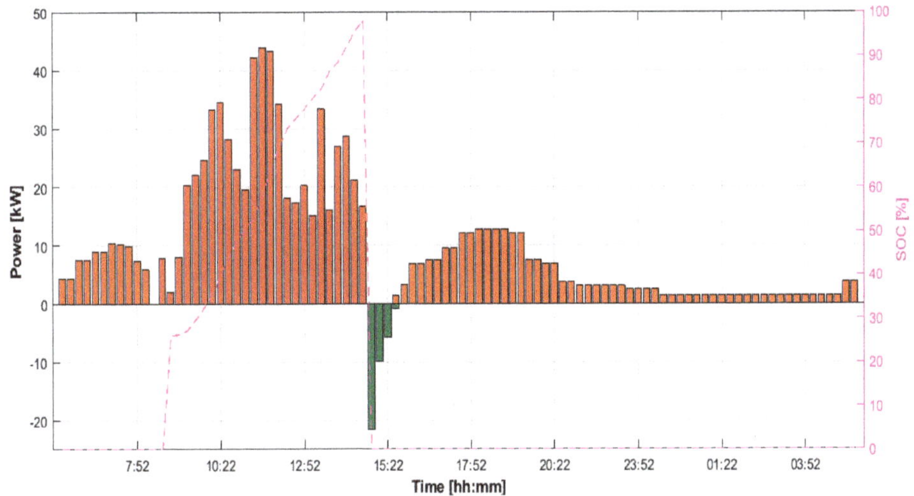

**Figure 19.** Final energy absorption of the car park distribution.

The green curve represents instead the residual photovoltaic energy that is still usable. From the producibility analysis carried out in the other months, it is also shown that the addition of the new electric load still leaves a good margin of unused power, which can thus be transferred to the network, for example, through the net-metering.

Alternatively, the design of a suitable energy storage system can also be assumed. In January for example, photovoltaic production provided 75% of energy needs. Instead, if we consider the most productive month, the estimated average daily reducibility would be about 0.77 MWh. If the energy needed to feed the parking lot and the EVs is subtracted, the portion of energy that can be exchanged with the network would be equal to 0.58 MWh. This amounts to nearly three times the energy actually needed.

### 6.2. Peak Shaving Algorithm

A different perspective to view or appraise EVs is to consider them not as a passive load, but as possible active elements of the electric network in order to explore their storage capacities. Implementation of a possible "peak shaving" algorithm is attempted in this paragraph. To do this, the energy stored in the batteries of cars connected to the EVSEs will be exploited. Based on the value of the power required by the car park's utilities, the controller will make an evaluation of whether to take the energy required from the network, or extract it from the vehicles.

In order to have a clearer idea of the applied operating principle, let us consider the graph shown in Figure 20.

The red dashed line represents the maximum absorbable power before the protections are activated. The two dotted lines in blue, Pref1 and Pref2, represent the two control values of the algorithm. In particular, if the energy demand is higher than Pref1, then the energy is taken from the car rather than from the grid. If the requested power is below Pref1, then the cars are loaded, on the whole, with a power equal to Pref2. Naturally, the nominal power of the supply equipment must always be respected.

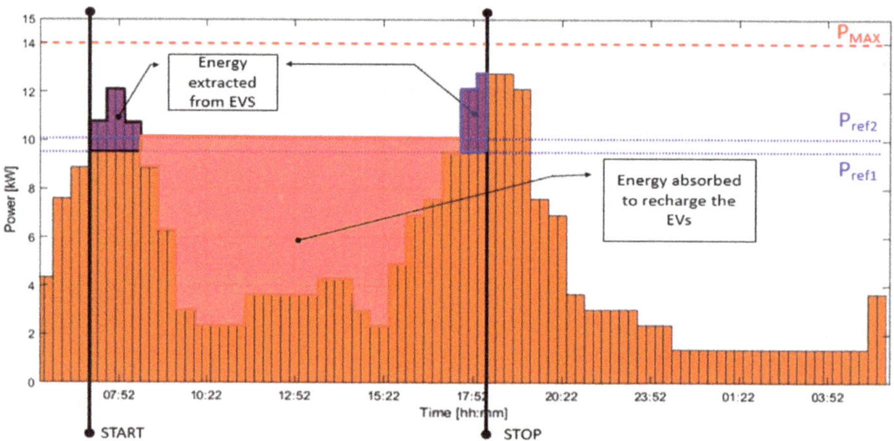


**Figure 20.** Operating principle of the peak shaving algorithm.

In Figure 21, it is possible to appreciate the final result of this algorithm. The dotted grey line represents the old load curve, while the red one shows the new load curve after the "peak shaving" algorithm is applied. The blocks in green represent the energy extracted from EVs to feed the parking in the phases of greater absorption, while the dotted blue line indicates the trend over time of the fleet's State Of Charge (SOC). As noted, through this algorithm, the charging of EVs can be extended for up to 10 or 11 h. For conclusive purposes, the flowchart related to the "peak shaving" algorithm is presented in Figure 22.

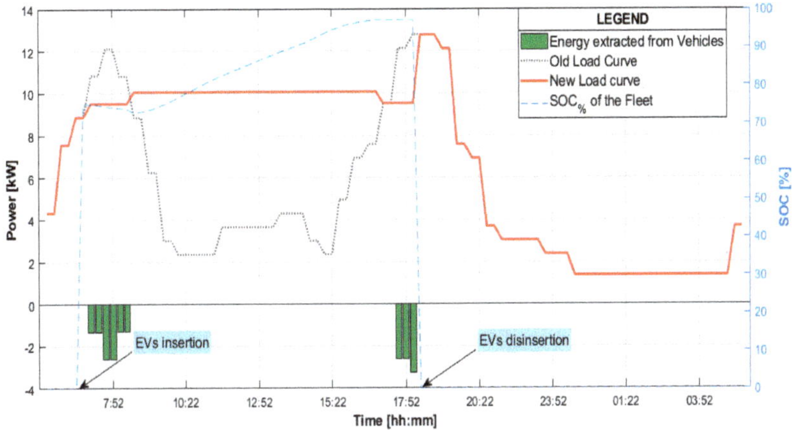


**Figure 21.** Comparison between the old load curve and the new load curve.

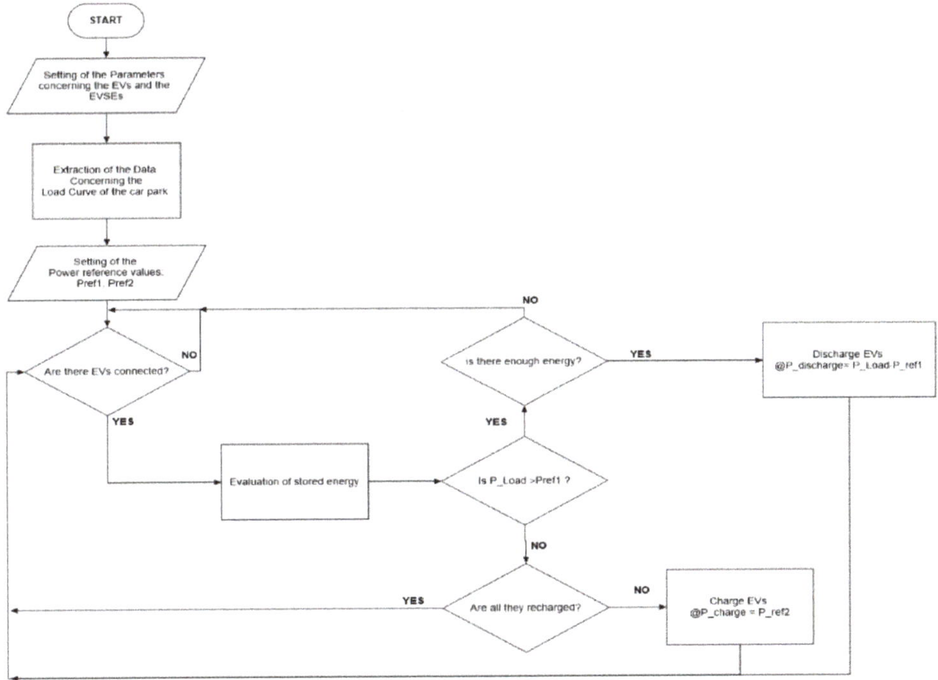


**Figure 22.** Flowchart of the "peak shaving" algorithm.

## 7. Conclusions

During this study, an attempt was made to outline the current status and possible future trends regarding electric mobility in the most complete and concise way possible. To date, the overall use of electric vehicles is still modest, although, as we have seen, the market is likely to expand in the coming years. The increasing frequency of extreme meteorological phenomena related to climate change due to greenhouse gases, and the continuous reduction in fossil fuel deposits make it not only necessary, but a duty to migrate to a more sustainable means of transportation.

However, the unsupervised use of EVs on the roads could lead to dangerous overloads and sudden voltage drops for the distribution network, which would increase the probability of disservice. It is therefore true that, in parallel with their diffusion, it is necessary to envisage the implementation of an appropriate scheduling of the recharging process through intelligent recharging systems that are able to predict and analyze the situations of the distribution network and act accordingly.

Moreover, every new EV that is introduced into circulation also means a new load that needs electricity to recharge. Consequently, careful planning of renewable energy plants and their diffusion is also necessary in order to provide (in a manner as sustainably as possible) the huge amount of energy that these new "wheeled loads" will need.

In the second part of this study, we saw a practical application of this proposition. The electrical producibility of a photovoltaic system designed specifically for the car park of Ferrara railway station was analyzed and it was observed that not only was this plant able to supply energy for the basic load of the parking, but it also met the energy requirement for recharging the EVs. Furthermore, it leaves an additional residual energy that may be exploited through on-site exchanges. In conclusion, it may be stated that the biggest challenges facing V2G technology are those outlined below:

- From a technical point of view, the evolution of smart grids has yet to prove their robustness and reliability.
- From a technical perspective, the evolution of smart grids are yet to prove their robustness and reliability.
- From a regulatory perspective, an appropriate regulatory framework has to be defined in order to regulate the introduction of this new "active" load.
- From an economic viewpoint, in order to explore all the possible monetary advantages linked to this new paradigm, appropriate business plans need to be written alongside marketing campaigns that encourage the end user to connect their vehicles and enable operators to perform such activities described.
- Finally, from a political perspective, public administration ought to be sensitive to a change of direction towards more sustainable modes of transport. It will be very important to encourage and support these technologies, and provide appropriate planning in terms of both logistics and infrastructure.

**Author Contributions:** R.G. and M.L. proposed the core idea and developed the models. R.G. and M.L. performed the simulations, exported the results, and analyzed the data. W.Y. and F.F. revised the paper. M.L. and W.Y. contributed to the design of the models and the writing of the manuscript. All authors have read and agreed to the published version of the manuscript.

**Funding:** This research received no external funding.

**Conflicts of Interest:** The authors declare no conflict of interest.

## References

1.  IEA/OECD. Global EV Outlook 2018: Towards Cross-Modal Electrification. International Energy Agency. Available online: https://webstore.iea.org/download/direct/1045?filename=global_ev_outlook_2018.pdf (accessed on 29 May 2020).
2.  IEA/OECD. 3030 Campaign. International Energy Agency. Available online: https://www.iea.org/media/topics/transport/3030CampaignDocumentFinal.pdf (accessed on 29 May 2020).
3.  Ramos, G.A.; Rios, M.A. Connection Schemes for Electric Vehicle Supply Equipment Retrieved. 2012, pp. 1–2. Available online: https://ieeexplore.ieee.org/document/6955230 (accessed on 29 May 2020).
4.  Qian, K.; Zhou, C.; Allan, M.; Yuan, Y. Modeling of Load Demand Due to EV Battery Charging in Distribution Systems. *IEEE Trans. Power Syst.* **2010**, *26*, 4–9. [CrossRef]
5.  Monteiro, V.; Pinto, J.; Afonso, J.L. Operation Modes for the Electric Vehicle in Smart Grids and Smart Homes: Present and Proposed Modes. *IEEE Trans. Veh. Technol.* **2016**, *65*, 1007–1020. [CrossRef]

6. Gungor, V.C.; Sahin, D.; Kocak, T.; Ergut, S.; Buccella, C.; Cecati, C.; Hancke, G.P. Smart grid and smart homes, Key players and pilot projects. *IEEE Ind. Electron. Mag.* **2012**, *6*, 18–34. [CrossRef]

7. Liu, C.; Chau, K.T.; Wu, D.; Gao, S. Opportunities and challenges of vehicle-to-home, vehicle-to-vehicle, and vehicle-to-grid technologies. *Proc. IEEE* **2013**, *101*, 2409–2427. [CrossRef]

8. Habib, S.; Kamran, M. A Novel Vehicle-to-Grid Technology with Constraint Analysis-A Review. In Proceedings of the International Conference on Emerging Technologies (ICET), Islamabad, Pakistan, 8–9 December 2014.

9. Uddin, K.; Dubarry, M.; Glick, M.B. The viability of vehicle-to-grid operations from a battery technology and policy perspective. *Energy Policy* **2018**, *113*, 342–347. [CrossRef]

10. Berger, L.; Iniewski, K. *Smart Grid, Applications, Communications and Security*; Wiley: Hoboken, NJ, USA, 2012.

11. Binding, C.; Gatenbain, D.; Jansen, B.; Sundostroem, O.; Andersen, P.; Marra, F.; Træholt, C. Electric Vehicle Fleet Integration in the Danish EDISON Project. Presented at 2010 IEEE Power & Energy Society General Meeting, Providence, RI, USA, 25–29 July 2010. [CrossRef]

12. Østergaard, J.; Foosnæs, A.; Xu, Z.; Mondorf, T.; Andersen, C.; Holthusen, S.; Holm, T.; Bendtsen, M.; Behnke, K. Electric Vehicles in Power Systems with 50% Wind Power Penetration: The Danish Case and the EDISON programme. Presented at European Conference on Electricity & Mobility, Wurzbug, Germany, 26–30 July 2009.

13. Surender Reddy, S.; Bijwe, P.; Abhyankar, A. Real-Time Economic Dispatch Considering Renewable Power Generation Variability and Uncertainty over Scheduling Period. *IEEE Syst. J.* **2015**, *9*, 1440–1451. [CrossRef]

14. Khan, S.U.; Mehmood, K.K.; Haider, Z.M.; Bukhari, S.B.A.; Lee, S.-J.; Rafique, M.K.; Kim, C.-H. Energy Management Scheme for an EV Smart Charger V2G/G2V Application with an EV Power Allocation Technique and Voltage Regulation. *Appl. Sci.* **2018**, *8*, 648. [CrossRef]

15. De Carne, G.; Zou, Z.; Buticchi, G.; Liserre, M.; Vournas, C. Overload Control in Smart Transformer-Fed Grid. *Appl. Sci.* **2017**, *7*, 208. [CrossRef]

16. Haider, Z.M.; Mehmood, K.K.; Rafique, M.K.; Khan, S.U.; Lee, S.J.; Kim, C.H. Water-filling algorithm based approach for management of responsive residential loads. *J. Mod. Power Syst. Clean Energy* **2018**, *6*, 118–131. [CrossRef]

17. Jhala, K.; Natarajan, B.; Pahwa, A.; Erickson, L. Coordinated Electric Vehicle Charging for Commercial Parking Lot with Renewable Energy Sources. *Electr. Power Compon. Syst.* **2017**, *45*, 344–353. [CrossRef]

18. Mehmood, K.K.; Khan, S.U.; Lee, S.J.; Haider, Z.M.; Rafique, M.K.; Kim, C.H. Optimal sizing and allocation of battery energy storage systems with wind and solar power DGs in a distribution network for voltage regulation considering the lifespan of batteries. *IET Renew. Power Gener.* **2017**, *11*, 1305–1315. [CrossRef]

19. Khosrojerdi, F.; Taheri, S.; Taheri, H.; Pouresmaeil, E. Integration of Electric Vehicles into a Smart Power Grid: A Technical Review. In Proceedings of the IEEE Electrical Power and Energy Conference (EPEC), Ottawa, ON, Canada, 12–14 October 2016.

20. Knezović, K.; Martinenas, S.; Andersen, P.B.; Zecchino, A.; Marinelli, M. Enhancing the role of EVs in the grid: Field validation of multiple ancillary services provision. *IEEE Trans. Transp. Electrif.* **2017**, *3*, 201–209. [CrossRef]

21. Project, P. Parker | Parker-Project. 2016. Available online: http://parker-project.com/ (accessed on 4 May 2020).

22. Martinenas, S.; Knezovic, K.; Marinelli, M. Management of Power Quality Issues in Low Voltage Networks using Electric Vehicles: Experimental Validation. *IEEE Trans. Power Deliv.* **2017**, *32*, 971–979. [CrossRef]

23. Parker Project. World'srst Cross-Bran d V2G Demonstration Conducted in Denmark. 2017. Available online: http://parker-project.com/wp-content/uploads/2017/11/Worlds-rst-cross-brand-V2G-demonstration-conducted-in-Denmark.pdf (accessed on 4 May 2020).

24. Marinelli, M.; Martinenas, S.; Knezović, K.; Andersen, P.B. Validating a centralized approach to primary frequency control with series-produced electric vehicles. *J. Energy Storage* **2016**, *7*, 63–73. [CrossRef]

25. Rezkalla, M.; Zecchino, A.; Martinenas, S.; Prostejovsky, A.M.; Marinelli, M. Comparison between Synthetic Inertia and Fast Frequency Containment Control on Single Phase EVs in a Microgrid. *Appl. Energy* **2018**, *210*, 764–775. [CrossRef]

26. Longo, M.; Lutz, N.; Daniel, L.; Zaninelli, D.; Pruckner, M. Towards an impact study of electric vehicles on the Italian electric power system using simulation techniques. In Proceedings of the 2017 IEEE 3rd International Forum on Research and Technologies for Society and Industry (RTSI), Modena, Italy, 11–13 September 2017.

27. Perujo, A.; Ciuffo, B. The introduction of electric vehicles in the private fleet: Potential impact on the electricity supply and on the environment. A case study for the Province Milan, Italy. *Energy Policy* **2010**, *38*, 4549–4561. [CrossRef]
28. Ekman, C.K. On the synergy between large electric vehicle fleet and high wind penetration—An analysis of the Danish case. *Renew. Energy* **2011**, *36*, 546–553. [CrossRef]
29. Hartmann, N.; Özdemir, E. Impact of different utilization scenarios of electric vehicles on the German grid in 2030. *J. Power Sources* **2011**, *196*, 2311–2318. [CrossRef]
30. Brenna, M.; Corradi, A.; Foiadelli, F.; Longo, M.; Yaici, W. Numerical simulation analysis of the impact of photovoltaic systems and energy storage technologies on centralised generation: A case study for Australia. *Int. J. Energy Environ. Eng.* **2020**, *11*, 9–31. [CrossRef]
31. Kempton, W.; Tomic, J. Vehicle-to-grid power implementation: From stabilising the grid to supporting large-scale renewable energy. *J. Power Sources* **2005**, *144*, 280–294. [CrossRef]
32. Tomic, J.; Kempton, W. Using fleets of electric-drive vehicles for grid support. *J. Power Sources* **2007**, *168*, 459–468. [CrossRef]
33. Sovacool, B.K.; Hirsh, R.F. Beyond batteries: An examination of the benefits and barriers to plug-in hybrid electric vehicles (PHEVs) and a vehicles-to-grid (V2G) transition. *Energy Policy* **2009**, *37*, 1095–1103. [CrossRef]
34. Zhang, Q.; Tezuka, T.; Ishihara, K.N.; Mclellan, B.C. Integration of PV power into future low-carbon electricity systems with EV and HP in Kansai Area, Japan. *Renew. Energy* **2012**, *44*, 99–108. [CrossRef]
35. Mwasilu, F.; Justo, J.J.; Kim, E.K.; Do, T.D.; Jung, J.W. Electric vehicles and smart grid interaction: A review on vehicle to grid and renewable energy sources integration. *Renew. Sustain. Energy Rev.* **2014**, *34*, 501–516. [CrossRef]
36. Drude, L.; Junior, L.C.P.; Rüther, R. Photovoltaics (PV) and electric vehicle-to-grid (V2G) strategies for peak demand reduction in urban regions in Brazil in a smart grid environment. *Renew. Energy* **2014**, *68*, 443–451. [CrossRef]
37. Habib, S.; Kamran, M.; Rashid, U. Impact analysis of vehicle-to-grid technology and charging strategies of electric vehicles on distribution networks—A review. *J. Power Sources* **2015**, *277*, 205–214. [CrossRef]
38. Van der Kam, M.; van Sark, W. Smart charging of electric vehicles with photovoltaic power and vehicle-to-grid technology in a microgrid; a case study. *Appl. Energy* **2015**, *152*, 20–30. [CrossRef]
39. Mozafar, M.R.; Amini, M.H.; Moradi, M.H. Innovative appraisement of smart grid operation considering large-scale integration of electric vehicles enabling V2G and G2V systems. *Electr. Power Syst. Res.* **2018**, *154*, 245–256. [CrossRef]
40. Uddin, K.; Jackson, T.; Widanage, W.D.; Chouchelamane, G.; Marco, J. On the possibility of extending the lifetime of lithium-ion batteries through optimal V2G facilitated by an integrated vehicle and smart-grid system. *Energy* **2017**, *13315*, 710–722. [CrossRef]
41. Das, H.S.; Rahman, M.M.; Li, S.; Tan, C.W. Electric vehicles standards, charging infrastructure, and impact on grid integration: A technological review. *Renew. Sustain. Energy Rev.* **2020**, *120*, 109618. [CrossRef]
42. Ioakimidis, C.S.; Thomas, D.; Rycerski, P.; Genikomsakis, K.N. Peak shaving and valley filling of power consumption profile in non-residential buildings using an electric vehicle parking lot. *Energy* **2018**, *1481*, 148–158. [CrossRef]
43. Zhang, L.; Jabbari, F.; Brown, T.; Samuelsen, S. Coordinating plug-in electric vehicle charging with electric grid: Valley filling and target load following. *J. Power Sources* **2014**, *267*, 584–597. [CrossRef]
44. Fazelpour, F.; Vafaeipour, M.; Rahbari, O.; Rosen, M.A. Intelligent optimization to integrate a plug-in hybrid electric vehicle smart parking lot with renewable energy resources and enhance grid characteristics. *Energy Convers. Manag.* **2014**, *77*, 250–261. [CrossRef]
45. Hota, A.R.; Juvvanapudi, M.; Bajpai, P. Issues and solution approaches in PHEV integration to smart grid. *Renew. Sustain. Energy Rev.* **2014**, *30*, 217–229. [CrossRef]
46. Tan, K.M.; Ramachandaramurthy, V.K.; Yong, J.Y. Integration of electric vehicles in smart grid: A review on vehicle to grid technologies and optimization techniques. *Renew. Sustain. Energy Rev.* **2016**, *53*, 720–732. [CrossRef]
47. Liu, J.; Zhang, T.; Zhu, J.; Ma, T. Allocation optimization of electric vehicle charging station (EVCS) considering with charging satisfaction and distributed renewables integration. *Energy* **2018**, *164*, 560–574. [CrossRef]

48. Hariri, A.M.; Hashemi-Dezaki, H.; Hejazi, M.A. A novel generalized analytical reliability assessment method of smart grids including renewable and non-renewable distributed generations and plug-in hybrid electric vehicles. *Reliab. Eng. Syst. Saf.* **2020**, *196*, 106746. [CrossRef]
49. Sultana, U.; Khairuddin, A.B.; Sultana, B.; Rasheed, N.; Qazi, S.H.; Malik, N.R. Placement and sizing of multiple distributed generation and battery swapping stations using grasshopper optimizer algorithm. *Energy* **2018**, *165*, 408–421. [CrossRef]
50. Reuters. Nissan Leaf Gets Approval for Vehicle-To-Grid Use in Germany. 23 October 2018. Available online: https://www.reuters.com/article/us-autos-electricity-germany/nissan-leaf-approved-for-vehicle-to-grid-use-in-germany-idUSKCN1MX1AH (accessed on 26 May 2019).
51. Nissan Leaf 2018. Available online: https://www-europe.nissan-cdn.net/content/dam/Nissan/gb/brochures/Vehicles/Nissan_Leaf_UK.pdfgermany/nissan-leaf-approved-for-vehicle-to-grid-use-in-germany-idUSKCN1MX1AH (accessed on 26 May 2019).
52. Robalino, D.M.; Kumar, G.; Uzoechi, L.; Chukwu, U.C.; Mahajan, S.M. Design of a Docking Station for Solar Charged electric and Fuel Cell Vehicles. In Proceedings of the International Conference on Clean Electrical Power (ICCEP), Capri, Italy, 9–11 June 2009; pp. 655–660.
53. Chukwu, U.C.; Mahajan, S.M. V2G Parking lot with PV Rooftop for Capacity Enhancement of a Distribution System. *IEEE Trans. Sustain. Energy* **2014**, *5*, 119–127. [CrossRef]

*Article*

# Use of AMT Transformers and Distributed Storage Systems to Enhance Electrical Feeding Systems for Tramways

**Romano Giglioli, Giovanni Lutzemberger * and Luca Sani**

Department of Energy, Systems, Territory and Constructions, University of Pisa, 56122 Pisa, Italy;
romano.giglioli@unipi.it (R.G.); luca.sani@unipi.it (L.S.)
* Correspondence: giovanni.lutzemberger@unipi.it; Tel.: +39-050-221-7311

Received: 18 August 2020; Accepted: 8 September 2020; Published: 10 September 2020

**Abstract:** Tramway systems are more and more diffused today, to reduce pollution and greenhouse emissions. However, their electrical feeding substations can have significant margin for improvement. Therefore, it is questionable which kind of changes can be introduced, by changing their main features. First of all, transformer technology can be enhanced, by moving from the standard transformer to the amorphous metal one; thus, guaranteeing a significant reduction in losses. Then, by installing one dedicated storage systems for each substation. This solution can help to increase the energy efficiency; thus, recovering the tram braking energy and reducing the delivered energy from the grid, and also the reliability of the system; thus, guarantee different levels of services, in the case of failure of a feeding substation. This paper investigates in a systematic approach the two proposed solutions. In particular an amorphous metal transformer has been properly designed, and performance compared to the standard one. Then, evaluation of distributed storage installation was performed, and the aspects of reliability for these systems evaluated. Results have shown the general feasibility of the proposed solutions, showing a significant energy saving with respect to the conventional ones.

**Keywords:** amorphous transformer; energy storage; failure; feeding substation; tramway

---

## 1. Introduction

Tramway systems are more and more diffused today, in order to limit environmental impact from transportation field. However, their electrical feeding substations show significant margin for improvements by introduction of new features, to improve their technical characteristics.

Firstly, feeding system solutions typically do not allow the reversibility of power fluxes from electrical feeding substations (ESSs). Therefore, the braking energy can be recovered only by other trains running, when available to do that (e.g., during acceleration or constant speed phases). As deeply investigated in [1–4], if some storage capability is used, the amount of the energy to recover can be significantly enhanced.

Another option typically considered in the literature is the use of bidirectional feeding substations without any storage capability, as described in [5]. However, we must consider that in this case the amount of the recoverable energy is slightly reduced, since losses increase due to the high distances to be covered. Additionally, the absence of storage capability does not help in improving the system reliability, and the overall cost of the system is very high due to its complexity.

If non-reversible feeding substations equipped with storage are considered, the authors already did the calculation of the energy saving derived from introduction of some stationary storage systems on an existing tramway, through a simulation tool specifically developed in Modelica language, which include the electrical network, the vehicles, and the driver [6]. This tool has been validated through the experimental measurements collected from an existing tramway and outputs crosschecked also with results given by different tools, in different case studies [7]. Once developed and properly set, the tool has allowed to demonstrate, from the evaluation of the annual demand of electricity from electrical feeding substations (ESSs), the cost-effectiveness due the introduction of a limited number of storage systems, allowing a payback time shorter with respect to the useful plant life [8,9]. However, it can be of interest to analyze if the number of installed storages can further increase with respect to what already studied, up to the installation of one storage for each electrical feeding substation (ESS), in order to increase the system reliability and to guarantee adequate levels of redundancy in the case of failure of one or more electrical feeding substations (ESSs). Additionally, when they are installed on each substation, they can also reduce the size of the transformers, since they are able to sustain peaks of power requested by the different trains on track. These aspects will be widely analyzed in this paper.

Additionally, the extensive campaign of measurements also has shown the high amount of losses caused by the transformers linked to the MV network. Starting from this, a deep analysis about component technologies inside each electrical feeding substation (ESS) was carried out. In particular, utilization of the amorphous core transformer (AMT) was considered, by designing a customized version for the considered application, in order to improve efficiency and thus verifying how much losses can be reduced.

This paper therefore shows how to systematically improve the electrical feeding substations (ESSs) from the point of view of the electrical energy utilization, moving from the improvement of the transformer technology and of the solutions regarding stationary storage systems, within a cost-effectiveness perspective.

## 2. Technology Improvements

### 2.1. Transformer

In a standard electrical feeding substation (ESS), a three-phase distribution transformer with two LV windings, star, and delta, and a MV delta winding, is typically used. It is manufactured so there is no phase displacement between the MV secondary winding and the LV delta winding and hence the star LV secondary winding leads the MV winding by 30°. Each LV winding is then connected to a converter's bridge [10].

This transformer typology, as all electrical machines, has losses in the windings (in losses), due to the current circulation, and in the iron (no-load losses) due to magnetic hysteresis and eddy currents. Although the latter are a small fraction, their contribution in energy is not negligible compared to load losses because the transformer is always connected to the network and then subjected to the grid voltage almost constant over time: thus their no load losses remain unchanged. Except for densely populated urban centers, the average power delivered by a MV/LV transformer, during a week, is less than 40% of the nominal power. This value is reduced to 10% at night. A cabin transformer generally is oversized to compensate for, in the future, any increase of the output power, taking into account that the average useful life exceeds thirty years. Load losses, by following the same trend, are therefore always lower than the nominal one. All this explains the interest in industry and regulations, regarding the realization of machines with lower iron losses. In the joint overlap regions, the losses increase around 50% due to interlaminar flux and deviation of flux from rolling direction.

In the case of converter transformer, under normal operating conditions, the current in the windings has a stepped waveform. For a 12-pulse operating condition, the converters create harmonic of order $12k \pm 1$, where k is integer, at the HV side. The presence of harmonics is one of the most severe aspect of the converter transformer. Harmonics increase eddy losses in the windings and stray losses in structural elements of the converter transformer. In addition, due the asymmetries in valve firing angles in the converter, the currents in the transformer windings have a DC component. In order to minimize excessive losses and noise, an accurate design of the magnetic circuit is needed.

Transformers distribution (with the high voltage (HV) less than 24 kV and oil cooling) are classified according to IEC 50464-1 based on the rated power and divided into subclasses according to the value of no-load losses (five classes from $A_0$ to $E_0$) and load losses (four classes from $A_k$ to $D_k$). The state of the art is represented by machines in class $C_k A_0$.

Regulation 548/2014 of the European directive in 2009 defined the minimum efficiency requirements and deadlines for new distribution transformers placed on the European market: class $A_k A_0$ from July 2021 [11]. Currently in Europe, the power transformers core is made with regular grain-oriented (RGO) silicon steel. This technology is sufficiently mature so that the achievement of greater efficiency classes may be obtained only by decreasing the current density in the coils and the magnetic induction in the cores, increasing the volume of the transformer. Furthermore, the production process has a low degree of automation. Considering this background, the request for high efficiency transformers could increase significantly in the future; therefore, the excellent market perspective for these machines is evident.

A really interesting solution is the amorphous metal transformer (AMT). It was introduced in the US market (in the mid-1980s), and in Southeast Asia (Japan and China) during the last decade. The amorphous metal is a metallic alloy of iron, boron, and silicon (Fe-B-Si) made by solidifying alloy melts at rates rapid enough to prevent the metal crystallization [12,13]. Such rapid solidification leaves a vitrified solid with a random (amorphous) atomic structure. The high heat extraction rates constrain the solid in the form of a thin ribbon, about 25 μm thick. Due to the presence of boron, amorphous metal has a reduced saturation induction than RGO steel. As result, amorphous core transformers often have a larger core cross-sectional area, resulting in larger coils and transformer footprint. The most significant characteristic of an amorphous metal in a transformer is that it yields a much lower core loss than even the best grades of RGO steel, by up to 70% [14].

Since the amorphous ribbon does not have a mechanical stiffness, the transformer core no longer has the function of holding up the coils. As a result, the structural design of this machine is significantly different compared to a transformer with RGO silicon steels, as shown in Figure 1a.

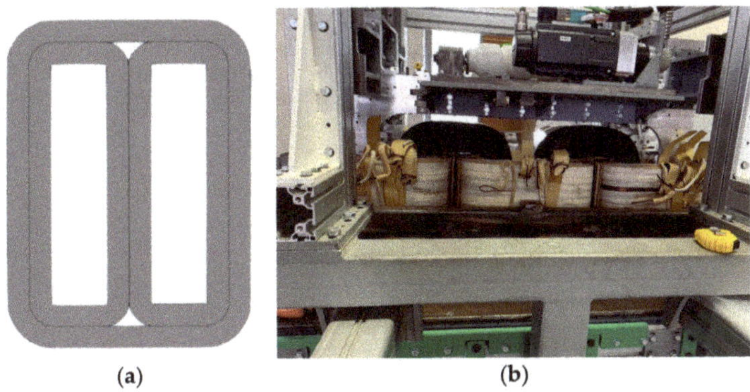

(a)          (b)

**Figure 1.** Three-legged transformer core made by amorphous alloy (**a**); transformer continuous core [15] during the automated manufacturing process (**b**).

Due to their significant different mechanical properties, another transformer manufacturing technique is also needed. Since the material is thin, the application of amorphous metal is restricted to wound transformer cores. Until now, the solution has been a wound core with distributed gaps: the ends of each ribbon are lapped with each other. The flux concentrations in the neighborhood of the joint gaps cause additional iron losses.

By exploiting the mechanical characteristics (flexibility) of ribbon [15] proposes an innovative assembly process for wound ferromagnetic core in which the joint gaps are eliminated, as in Figure 1b. All this, in addition to further improvement of the efficiency, will ensure a high standard product.

The technological innovation consists in being able to wind up a ribbon of magnetic material directly into the preformed coils. In particular the originality consists in providing a removable automated system, constituted by a series of rolls, on which the thin sheet of amorphous material is wound (with a thin layer of thermosetting resin deposit), mounted on a motorized guide with an optimized shape. The shape is a closed loop (rectangular with rounded corners or elliptical) formed around the coils.

The production process is completely different from the one currently used for the production of core laminations, and the solution developed will allow to introduce a high degree of automation, so that the productivity of its plants can be significantly increased.

Typically, an AMT is always more expensive than a silicon steel unit but can be more economical in many power systems. To specify cost-effective transformer performance, utility engineers commonly use a "loss-evaluation" method. This approach considers transformer loading patterns, energy costs, inflation, interest rates, and other economic factors to calculate the net present value of a watt of electric power. The combination of the initial cost of the transformer with its cost of operation, is summarized by the total owning cost (*TOC*) [16]:

$$TOC = B_P + (A \cdot C_L) + (B \cdot L_L) \tag{1}$$

where $B_P$ is the bid price, $A$ is the core loss factor, $C_L$ are the core losses (e.g., losses occurring in the magnetic core, due to alternating magnetization), $B$ is the Load Loss Factor, and $L_L$ are the load losses (e.g., losses when load currents flow). In the case of energy costs sufficiently high, AMTs make economic as well as environmental sense.

### 2.2. Storage System

As said, a train can send its braking energy on the catenary, only in the vicinity of other trains running, available to adsorb that energy. Therefore, when a stationary storage system is introduced, energy recovery is enhanced. In this way, also when no other trains are present, the storage system can adsorb the energy from the trains engaged in braking, and delivering it at a different time, in the presence of enough load. A detailed description of the problem, together with evaluation of the energy saving from electrical feeding substations, is widely described in [6,8,9]. The storage is normally not directly linked to the grid, but it is interfaced through the use of a DC/DC converter, having different functionalities: first of all, it is possible to control the energy flows; thus, preserving the storage system by the delivery or adsorption of high current peaks. Then, SOC drift can be avoided; therefore, maintaining the battery at an intermediate SOC value; thus, avoiding progressive charging or discharging, leaving it able to recover or deliver energy. Finally, to guarantee flexibility in the storage sizing, having the battery voltage not dependent from the operating pantograph voltage.

At least in theory, each electrical feeding substation (ESS) can be equipped with its dedicated storage or, to reduce costs, just a few storage systems can be installed, in correspondence to one or a few more substations. Certainly, when the number of storage systems equals the number of the feeding substations (ESSs), it is possible to reduce the sizing of the transformer, e.g., up to half of the original power, since the extra-power needed can be delivered by the storage systems.

This aspect, also by guaranteeing a reduction of the transformer *TOC*, as general rule, can therefore compensate, or at least cancel, the extra-costs needed for the storage systems. Additionally, extra-services can also be provided, by guaranteeing the train are at least able to reach the nearest train stop, in the case of failures of one or more ESSs. In this way, system reliability is enhanced, and adequate levels of redundancy are given.

Another aspect of interest is given by the technology used for the application. In fact, many lithium-based technologies are today available. One of the most common, based on utilization of lithium-iron-phosphate (LFP) cells [17], considered also in [8,9], which are typically considered in energy-oriented applications, and characterized by very low costs. In the last years, other technologies have been more and more considered, much more oriented in delivering or adsorbing high powers for short time durations. In particular lithium-titanate (LTO) cells [18] are specifically power-oriented, and therefore aligned to the application requirements, in which high current peaks have to be adsorbed during the regenerative braking of the trains. In fact, according to authors experience and manufacturer indications [18], they are able to sustain high charging or discharging current peaks, in the order of ten times their nominal capacity, for several tens of thousands of cycles; thus, showing also a significant life resistance to number of cycles.

## 3. The System Under Study

### 3.1. Architecture

The system under analysis is in Bergamo (Italy). The pattern has a full length of 12 km, and it is characterized by 10 feeding electrical substations (ESSs), whose positions are in Figure 2.

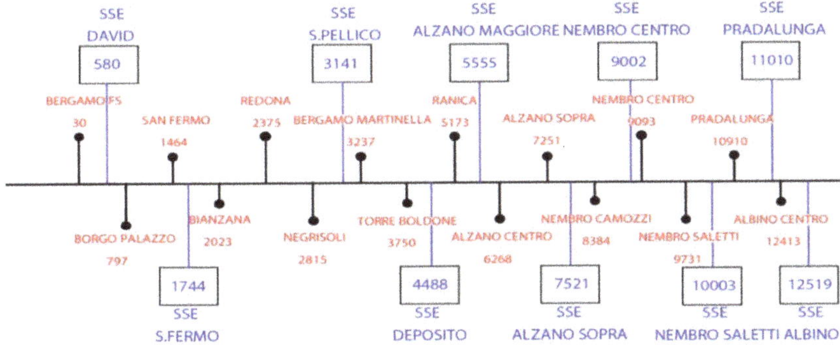


**Figure 2.** Position of the electrical feeding substations (ESSs) in blue and stops in red along the path (distances from the terminal are indicated in meters).

The main technical data of this system are in the next Table 1.

**Table 1.** Main features of the system.

| Tram | |
| --- | --- |
| **Empty Mass (t)** | 41.9 |
| Load mass (t) | 17.4 |
| Max power (kW) | 628 |
| Max tractive force (kN) | 72.1 |
| Max speed (km/h) | 70 |
| **Electrical Feeding Substations** | |
| Medium voltage (kV) | 15 |
| Output No load DC voltage (V) | 798 |
| Max power (MW) | 2.0 |

The feeding supply system is schematically represented in the next Figure 3. How visible, each ESS has a MV/LV transformer (TR1) with its converter, in order to feed energy needed for propulsion. Then, a second transformer (TR2) feeds the LV auxiliary loads, e.g., visual and acoustic signals of each ESS and of the nearest train stop. As always depicted in Figure 3, energy meters were installed in each ESS to measure energy flows, and also additional data logger were introduced, to measure instantaneous DC voltage and current. Measurement devices are also indicated in Figure 3.

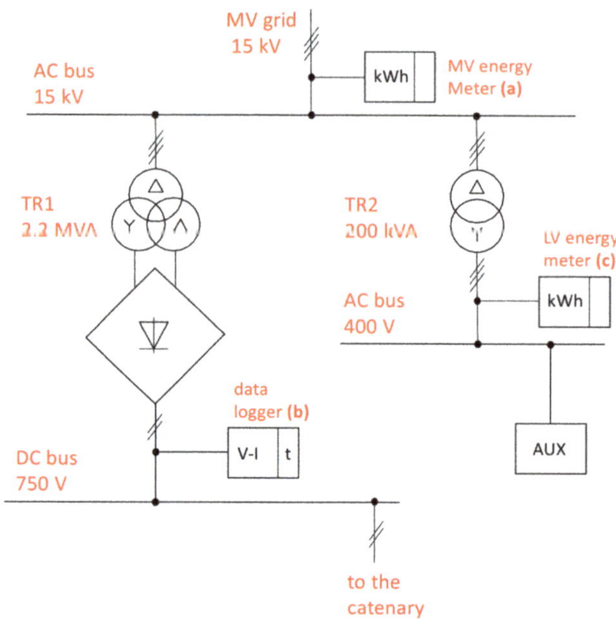



**Figure 3.** Scheme of ESS, including measurement system.

The main characteristics of the transformer TR1 are in the next Table 2.

**Table 2.** Main parameters of the standard transformer (TR1).

| Rated Power | 2.2 MVA |
| --- | --- |
| Voltage (primary/secondary) | 15/04 kV |
| Number of primary windings | 1 (Delta) |
| Number of secondary windings<br>Vector group | 2 (Star–Delta)<br>Dy11d0 |
| Medium | Dry |
| Iron core Layout<br>Iron core Material | Three limbs<br>Cold-rolled grain-oriented steel (CRGO) |
| Winding material | Copper |
| Maximum load losses $P_k$ | 17.5 kW |
| Maximum no-load losses $P_0$ | 4.5 kW |
| Efficiency at full load | 0.990 |

## 3.2. Energy Flows

Campaign of measurements accurately described in [8] allows to evaluate how energy fluxes are divided between losses of the devices in each ESS, energy demands of the trains, e.g., the energy needed for traction and for on-board auxiliary loads, and the energy required by LV auxiliary loads. Moreover, these measurements have been performed considering the real number of trains in operation, variable depending on the considered day. In this way, school working days, having some hours per day characterized by ten trains in operation, no-school working days, in which five trains are in operation, and finally holidays, having just three trains on track. Additionally, also effects of the season were considered, e.g., having different contributions from heating or air conditioning systems.

Results for a mid-season no-school working day, in the case of three ESSs, are in the next Table 3. Energy losses are in the range 20–24%. Further details are in [8,9].

**Table 3.** Daily energy flows for three ESSs.

| ESSs | Daily Energy MV (kWh) | Aux Energy LV (%) [1] | ESS Losses (%) [1] | Traction Energy (%) [1] |
|---|---|---|---|---|
| San Fermo | 1231 | 16.3 | 18.4 | 65.3 |
| Alzano | 882 | 22.3 | 22.4 | 55.3 |
| Albino | 1167 | 10.8 | 18.5 | 70.7 |

[1] Related to the corresponding daily MV energy.

As further step, yearly consumptions have been measured, and shown in the next Table 4. The full energy consumption from MV is shared between energy demand for traction, losses and LV auxiliary loads.

**Table 4.** Annual energy demand.

| | Energy from ESSs (MWh/y) |
|---|---|
| Total MV Energy (a) | 4100 |
| Traction (b) | 2756 |
| LV Aux Energy (c) | 738 |
| Losses | 606 |

How visible, annual energy losses from the transformer and its dedicated converter are about 15% of the total MV energy demand. It is therefore needed an accurate in-depth analysis of the standard transformer losses, thus evaluating the benefits given by utilization of the amorphous core transformer technology (AMT).

## 4. The Amorphous Metal Transformer (AMT)

### 4.1. Design of the Transformer

In order to improve the system reliability, the transformer TR1 can be replaced by two identical transformers having half power (i.e., 1100 kVA), in parallel connection. The main aim of the activity was therefore the design of a 1100 kVA AMT transformer, having one primary and two secondary windings.

First of all, starting from the architecture depicted in Figure 1a, it has been made the analytical winding size, then moving to the CAD modelling and finally the FEM analysis. Main parameters of the 1100 kVA rectifier transformer with amorphous alloy core have been finally summarized in the next Table 5. The design is compliance with EN-60146–1-3. The rated voltage and currents in primary and secondary windings of the transformer are 15 kV/0.4 kV and 42 A/538 A, respectively.

The operating flux density in the transformer core is fixed around 1.6 T to mitigate the effects of harmonics and dc component of the winding current. Due to the shape of the amorphous alloy strip, the cross section of the core is generally made rectangular, and the winding is also made rectangular. Delta and star LV windings are insulated from one another. MV windings are wrapped between the LV windings. Usually, LV windings are foil wound and MV may be windings disc wound. Both are made by aluminum.

As remarked, the detailed 3D FEM analysis has been carried out in order to validate the design of the transformer by using Simcenter Magnet software [19].

**Table 5.** Main parameters of 1100 kVA amorphous core transformer (AMT) transformer.

| Rated Power | 1.1 MVA |
| --- | --- |
| Voltage (primary/secondary) | 15/04 kV |
| Number of primary windings | 1 (Delta) |
| Number of secondary windings | 2 (Star–Delta) |
| Vector group | Dy11d0 |
| Medium | Dry |
| Iron core Layout | Three limbs |
| Iron core Material | Amorphous Metglass 2600 |
| Weight | 2200 kg |
| W × H × D | 1100 mm × 1500 mm × 284 mm |
| Winding material | Aluminum |
| Core mass | 980 kg |
| Magnetic flux density | 1.4 T |
| Vcc % | 3.5 % |
| Maximum load losses $P_k$ | 6.0 kW |
| Maximum no-load losses $P_0$ | 0.6 kW |
| Efficiency at full load | 0.9924 |

### 4.2. Application to the Tramway System

After designing the AMT transformer, it is important to evaluate which help this latter can guarantee on the full amount of energy consumption evaluation, as in the previous Table 4. Tables 2 and 5 show that both load and no-load losses are reduced. In detail, load losses are decreased from 17.5 kW to 12 kW in the case of two AMT transformers connected in parallel, while a significant reduction concerns the no-load losses, moving from 4.5 kW to 1.2 kW. Although these values are negligible with respect to the load losses amount, we must consider that they occur on the full number of hours per day, while load losses are present only in the average period of use. This is because, it is therefore needed to calculate through a weighted average the effects of the contributions of losses. In particular, we can consider what is currently shown in Equation (2).

$$L_{tot} = n_t \cdot n_d \cdot (t_{CL} \cdot C_L + t_{LL} \cdot L_L) \tag{2}$$

where $L_{tot}$ is the full amount of losses per year given by utilization of standard transformers, $n_t$ is the number of the transformers, $n_d$ is the operative number of days per year, $C_L$ are the core power losses calculated for the time duration $t_{CL}$, i.e., equal to 24 h/day, $L_L$ are the full load losses, calculated for the average time duration $t_{LL}$. Moving from the previous experimental evaluation of losses, for which 606 MWh/y were measured (see Table 4), and by using data of standard transformers, i.e., having $C_L$ and $L_L$ respectively equal to 4.5 kW and 17 kW, it was then possible to obtain the average time duration $t_{LL}$ by utilization of Equation (2). It was finally considered 10 installed transformers, and 8760 operative hours per year. In particular, we obtained:

$$t_{LL} = \frac{L_{tot}}{n_t \cdot n_d \cdot L_L} - \frac{t_{CL} \cdot C_L}{L_L} = 3.4 \, h/day \tag{3}$$

Moving from this, relation Equation (2) was newly adopted, by changing $C_L$ and $L_L$ to the values of the AMT transformers, e.g., 1.2 kW and 12 kW, respectively (see Table 4), and by taking unmodified the previously time durations. With the new sizing for the transformer, and by taking unmodified the architecture of Figure 3, losses changes from 606 MWh/y to 254 MWh/y; thus, having a 58% reduction.

Therefore, by taking the same requests for traction (e.g., 2756 MWh/y) and auxiliary loads (738 MWh/y), the total MV energy request is reduced to 3748 MWh/y.

In conclusion, the total MV energy shows an 8% reduction, and losses are about 7% of the total MV energy demand. The amount of losses are therefore considerably reduced compared to before. This is visible also from Equation (1). In fact, *TOC* reduction is about 20% than for the standard transformer.

It is finally questionable if also the TR2 transformer, aimed to feed auxiliary loads, could be replaced with the introduction of a second AMT transformer. However, its reduced power (i.e., 200 kVA with respect to 2.2 MVA) would make a negligible reduction in the LV aux energy spent, if an equivalent improvement of efficiency as for the TR1 would be considered, in the order of a few MW per year.

## 5. Storage System Integration

### 5.1. Storage System Sizing

The architecture of each ESS, when equipped with some storage capability, in the next Figure 4, while the final sizing of the considered stationary storage system is instead shown in Table 6. The first rows show the two different lithium-based technologies which have been considered. The first one is based on LFP cells [17], while the second one is based on LTO cells [18].

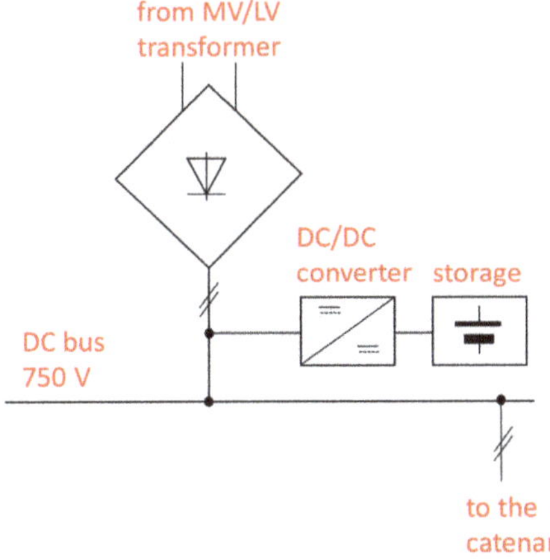


**Figure 4.** Scheme of the ESS equipped with storage system.

**Table 6.** Storage system sizing.

| Technology | LFP | LTO |
| --- | --- | --- |
| Nominal cell voltage (V) | 32 | 2.4 |
| Nominal cell capacity (Ah) | 160 | 10 |
| Max allowed charging current (A/Ah) | 6 | 60 |
| Cost (€/kWh) | 150 | 600 |
| Number of cells in series | 190 | 254 |
| Number of cells in parallel | 1 | 5 |
| Nominal battery capacity (Ah) | 160 | 50 |
| Nominal battery energy (kWh) | 97.3 | 30.5 |

According to manufacturer indications [18], the maximum allowed charging cell current is extremely higher for LTO, which value has been calculated from the declared input power, fixed at 1500 W. It must also be said that the LFP manufacturer [17] does not give precise indications regarding performance in the case of charging pulses, which have been prudently fixed to 6 A/Ah, with respect to

10 A/Ah declared for discharging pulses. Finally, cell costs have been estimated from experience by the authors [8].

The sizing of the battery pack was made having as reference the peak of charging current, which the battery is subjected during the train braking phases, i.e., about 1000 A with 10 trains on the line corresponding to a rush hour. How visible from the last rows of Table 6, the 160 Ah-capacity selected for the LFP cell perfectly matches this performance requirement. However, the capacity has to be selected also to guarantee a corresponding SOC variation compatible with the expected life of that battery. In this way, the number of charging-discharging cycles, having every day up to 10 trains on the line, which are continuously engaged in acceleration and braking phases, can reach hundreds of thousands in a few years. This solicitation can be sustained from the battery only when the corresponding SOC variation during cycling is within the range 5–10% [20–22]. In the case of the LFP solution, because of their worst performance, SOC variation under this condition must not overcome 5%. Regarding the LTO solution, although a smaller capacity would be acceptable as well regarding the maximum stress required, a capacity of at least 50 Ah is needed to stay within the indicated SOC variation, to preserve the battery life. Moreover, different lithium cell technologies (e.g., NMC, NCA, etc.) could at least be used in theory; however, as first approximation their general characteristics can be aligned to the typologies here described, and their performance within the considered range, mainly LTO oriented when able to sustain high charging/discharging current rates. An additional example describing the utilization of high-power batteries is detailed in [23].

The quantity of the braking energy to be recovered in the different scenarios was analyzed by means of a simulation tool developed in Modelica language [24] in order to evaluate the power and energy flows on the overall system. In this way, the supply network, the catenary, and the trains running have been simulated, following the same approach as in [6,25].

The tool was then validated by the utilization of the daily and annual energy flows, respectively shown in Tables 3 and 4. Obviously, daily energy flows have made it possible to correctly calibrate the ESS model parameters, the braking energy control strategy and the trains energy consumptions, by setting the maximum allowed speed along the different sections of the route, and the different mix of auxiliary loads. In this way, the tool was able to be aligned with the indicated energy required for traction in the last row of Table 4, with a gap of about 1%.

Once validated, the tool was to evaluate the energy saving gained from the installation of storage systems, from one up to ten, e.g., one for each ESS. Energy saving was evaluated as difference between the total energy delivered by the ESSs in the configuration without storage, and the total energy delivered by the ESSs in the configuration with the number of the storage systems examined. Results, taken from [9] for clarity, are in the next Table 7.

**Table 7.** ESSs energy demand for traction and related energy saving, for different storage systems option.

| Case | Energy Delivered by the ESSs (MWh/y) | Energy Saving (MWh/y) |
|---|---|---|
| Base (no storage) | 2756 | – |
| 1 Storage system | 2604 | 180 |
| 3 Storage systems | 2088 | 696 |
| 5 Storage systems | 1805 | 979 |
| 10 Storage systems | 1680 | 1104 |

It is apparently questionable if the storage sizing needs to be updated when the number of storages rises up from one up to ten units. In fact, at equal number of trains, power, and energy flows to be managed would be shared in a larger number of storages; thus, allowing a downsizing in terms of nominal capacity. However, it is also possible to observe how the need to preserve the SOC fluctuation within a narrow band, makes the choice of maintaining the original storage sizing. This is visible also in the next Figure 5, where battery current and SOC are shown in the case of ten storage systems, for which LTO technology has been chosen. As said before, current peaks do not overcome 1000 A, and SOC variation is about 10%. Figure 5 refers to the case of a rush hour, with 10 trains running.

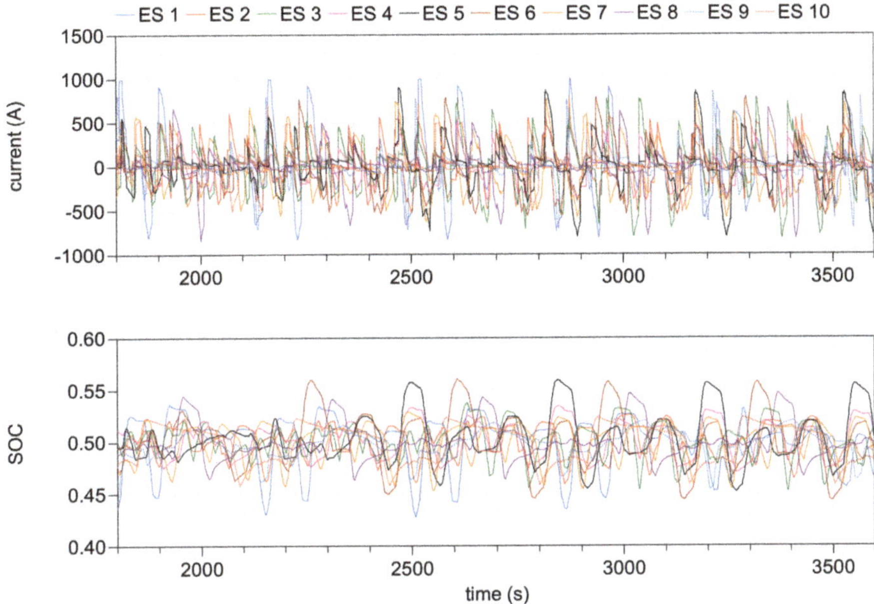


**Figure 5.** Battery current and SOC when having 10 storage systems installed, rush hour (ten trains running).

### 5.2. System Reliability

From the numbers of Table 7, it is clear that the best choice from the energy saving point of view is the installation of three storage systems; in fact, further energy saving due the increment of the storage systems up the three units cannot be compensated by their extra-purchase costs. Some sensitive analysis for different scenarios, including different costs of electricity and storage technologies are detailed in [8]. However, installation up to 10 storage systems, although less cost-efficient, may be useful as well in order to improve the system reliability. In this regard, two possible failures have to be considered:

The first failure typology takes into consideration that all the ESSs are subjected to blackout, with the electrical feeding from the grid immediately stopped. In this way, each storage system installed must serve one train on average, allowing it to reach at least two nearest subsequent train stops. This means, by considering the sequence of train stops of the previous Figure 2, each train has to cover under these conditions a traveled distance not higher than 3 km.

From simulation results, the traction energy consumption at constant speed is in the range 2.6–3.2 kWh/km for the trains running at 50 or 70 km/h respectively. Therefore, the two considered configurations were checked. In this way, considering a mid SOC level and running at the highest speed (i.e., 70 km/h), each battery pack can move at the maximum constant speed (i.e., 70 km/h) the train for about 4.8 km or 15.2 km, respectively. Therefore, the requirement to let each train running to the nearest station has been fully met.

As second failure, it has been considered the situation in which one single ESS stops working. The failure can affect or not the stationary storage, installed in parallel. In this case, no change in normal operation is allowed, to guarantee the business continuity for the service. Naturally, both the two storage configurations have been simulated. The matrix shown in the next Table 8 shows all the considered case studied, which simulate a failure at the beginning, in the middle or at the end of the pattern, in which electrical feeding substations (ESSs) are numbered from 1 (e.g., David, see Figure 2) to up 10 (e.g., Albino, see Figure 2). When the failure of the storage is included, it is indicated in the corresponding row, always in Table 8. Table 8 shows results from simulation outputs in terms of functionality, indicating when service interruption (-) or business continuity (x) occur, in correspondence with the hypothesized failures.

**Table 8.** Matrix of test cases, failure of one ESS including storage or not, for different storage systems option; business continuity (x), service interruption (-).

| Failure | LFP Storage | LTO Storage |
|---|:---:|:---:|
| ESS1 | x | x |
| ESS1; storage 1 | - | - |
| ESS5 | x | x |
| ESS5; storage 5 | x | x |
| ESS10 | x | X |
| ESS10; storage 10 | - | - |

In this way, the following main remarks can be achieved. First of all, when the failure involves both the ESS and the corresponding storage, the business continuity cannot be guaranteed in each condition, if the reference power of 1 MW is installed for each ESS. In fact, due occasional overlap of acceleration and running phases of multiple trains in the same portion of the track, power peaks delivered from the ESS can reach a maximum value nearly about 1.6 MW. This is visible in the plot of the next Figure 6, where the maximum power from ESS1 is shown, in the case of all ESSs perfectly working, or when a failure concerns the nearest ESS0. How visible, the maximum extra-peak to be supplied is about 400 kW, and it remains sustainable when the full power of 2.2 MW is yet available, i.e., with standard transformers, or with two AMT transformers installed in parallel (see Section 4.2).

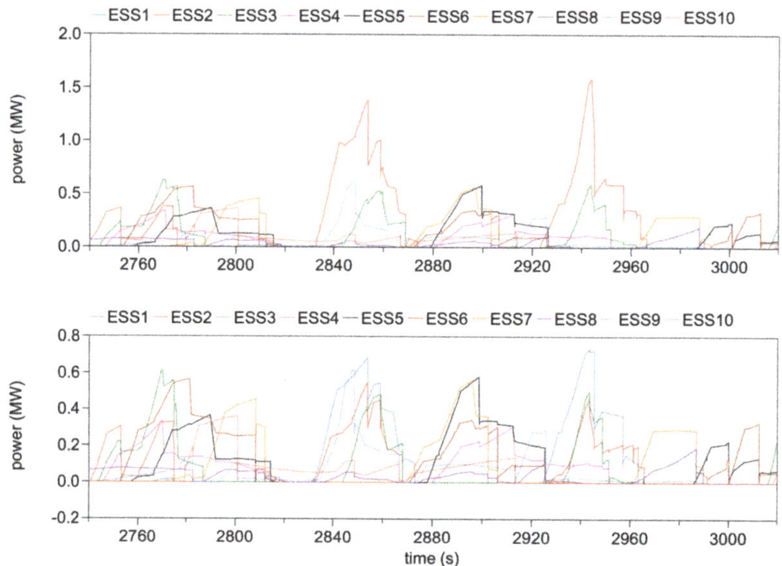

**Figure 6.** Power delivered from ESSs, in the case of failure from ESS1 (top) or no failure (bottom).

On the other hand, when half-power transformers having 1.1 MW are used, the extra-power needed has to be provided by the storage systems, and losses decrease from 4.5 kW and 17.5 kW (see Table 2), at exactly the half value, i.e., 2.3 kW and 8.8 kW, respectively; thus, further reducing the amount of losses calculated in the Section 4.2.

In particular, stationary storage systems are mandatory, in order to guarantee the extra-power needed. The two storage systems may appear useful as well, although the LTO configuration is properly designed for delivering high current rates, while the LFP configuration is exploited at its maximum performance, having about 600 kW delivered, aligned to the maximum allowed current rate of 6 A/Ah. Finally, when no overlap happens among multiple trains running, the business continuity is guaranteed also in the presence of a failure concerning both the storage and the feeding substation; this is the case when the failure refers to ESS4 and ESS9, always in Table 8.

## 6. Conclusions

This paper has described how it is possible to improve electrical feeding substations, by changing transformer technology and by installing dedicated high-power oriented storage systems.

In particular, a considerable reduction in losses up to 58% has been performed by simply changing the transformer technology, which guarantee a cost of ownership significantly lower than the standard transformer solution. Additionally, further losses reduction is obtainable by introducing half-power transformers.

Then, installation of storage systems was analyzed also by considering the improvement in terms of system reliability, in addition to energy efficiency. Results have shown also that installation of the storages is able to guarantee different tramway services, also when reducing the sizing of the installed transformers. Moreover, this last aspect can also compensate the extra-costs for the storages.

The presented results are produced on a specific case study, which represents a typical configuration for very common tramway systems and trolleybuses. Since significant similarities can be identified regarding the technical characteristics of these systems, it is therefore to be expected that the highlighted benefits from the proposed changes are nearly the same by moving from one system to another.

**Author Contributions:** Conceptualization, G.L., R.G., and L.S.; methodology, G.L., R.G., and L.S.; software, G.L. and L.S.; validation, G.L., R.G., and L.S.; formal analysis, G.L., R.G., and L.S.; investigation, G.L., R.G., and L.S.; resources, G.L., R.G., and L.S.; data curation, X.X.; writing—original draft preparation, G.L. and L.S.; writing—review and editing, G.L., R.G., and L.S.; visualization, G.L., R.G., and L.S.; supervision, R.G.; project administration, L.S.; funding acquisition, L.S. All authors have read and agreed to the published version of the manuscript.

**Funding:** This research was co-financed under Tuscany POR CREO 2014-2020 Project "FAT" (CUP 7165.24052017.112000040).

**Conflicts of Interest:** The authors declare no conflict of interest.

## Abbreviations

The following abbreviations are used in this manuscript:

| | |
|---|---|
| ESS | electrical feeding substation |
| AMT | amorphous core transformer |
| LV | low voltage |
| MV | medium voltage |
| RGO | gular grain-oriented |
| SOC | state of charge |
| TOC | total owning cost |
| LFP | lithium ferro-phosphate |
| LTO | lithium-titanate battery |
| NMC | nickel manganese cobalt oxide |
| NCA | lithium nickel cobalt aluminum oxide |

## References


1. Gonzalez-Gil, A.; Palacin, R.; Batty, P.; Powell, J. A systems approach to reduce urban rail energy consumption. *Energy Convers. Manag.* **2014**, *80*, 509–524. [CrossRef]
2. Gonzalez-Gil, A.; Palacin, R.; Batty, P. Sustainable urban rail systems: Strategies and technologies for optimal management of regenerative braking energy. *Energy Convers. Manag.* **2013**, *75*, 374–388. [CrossRef]
3. Lamedica, R.; Ruvio, A.; Galdi, V.; Graber, G.; Sforza, P.; Guidi Buffarini, G.; Spalvieri, C. Application of battery auxiliary substations in 3 kV railway systems. In Proceedings of the AEIT International Annual Conference (AEIT 2015), Naples, Italy, 14–16 October 2015; pp. 1–6.
4. Capasso, A.; Gannini, G.; Lamedica, R.; Ruvio, A. Eco-friendly urban transport systems. Comparison between energy demands of the trolleybus and tram systems. *Ing. Ferrov.* **2014**, *69*, 329–347.
5. Roch-Dupré, D.; Cucala, A.P.; Pecharromàn, R.R.; Lopez-Lopez, A.J.; Fernandez-Cardador, A. Evaluation of the impact that the traffic model used in railway electrical simulation has on the assessment of the installation of a Reversible Substation. *Int. J. Electr. Power Energy Syst.* **2018**, *102*, 201–210. [CrossRef]
6. Barsali, S.; Bolognesi, P.; Ceraolo, M.; Funaioli, M.; Lutzemberger, G. Cyber-Physical Modelling of Railroad Vehicle System using Modelica Simulation Language. In Proceedings of the Second International Conference on Railway Technology: Research, Development and Maintenance, Ajaccio, France, 8–11 April 2014; pp. 1–18.
7. Capasso, A.; Ceraolo, M.; Lamedica, R.; Lutzemberger, G.; Ruvio, A. Modelling and simulation of electric urban transportation systems with energy storage. In Proceedings of the IEEE 16th International Conference on Environment and Electrical Engineering (EEEIC), Florence, Italy, 7–10 June 2016; pp. 1–5.
8. Ceraolo, M.; Giglioli, R.; Lutzemberger, G.; Bechini, A. Cost effective storage for energy saving in feeding systems of tramways. In Proceedings of the IEEE International Electric Vehicle Conference (IEVC), Florence, Italy, 17–19 December 2014; pp. 1–6.
9. Fioriti, D.; Giglioli, R.; Lutzemberger, G.; Poli, D. Storage operation in tramway systems delivering grid services. In Proceedings of the IEEE 18th International Conference on Environment and Electrical Engineering (EEEIC), Palermo, Italy, 12–15 June 2018; pp. 1–6.
10. Harlow, J.H. *Electric Power Transformer Engineering*, 3rd ed.; CRC Press: Boca Raton, FL, USA, 2013; ISBN 9781439856291.
11. Steinmetz, T.; Cranganu-Cretu, B.; Smajic, J. Investigations of no-load and load losses in amorphous core dry-type transformers. In Proceedings of the he XIX International Conference on Electrical Machines (ICEM 2010), Rome, Italy, 6–8 September 2010; pp. 1–6.
12. Gatta, F.M.; Geri, A.; Maccioni, M.; Maiolini, C.; Paulucci, M.; Palone, F.; Cannavale, G.; Scaggiante, M. Impact of the recent EU regulation on HV/MV transformers: Possible technical-economic benefits arising from the replacement of old transformers. Two case studies. In Proceedings of the IEEE 16th International Conference on Environment and Electrical Engineering (EEEIC), Florence, Italy, 7–10 June 2016; pp. 1–6.
13. Carlen, M.; Xu, D.; Clausen, J.; Nunn, T.; Ramanan, V.R.; Getson, D.M. Ultra High Efficiency Distribution Transformers. In Proceedings of the IEEE PES Transmission and Distribution Conference and Exposition (IEEE PES T&D 2010), New Orleans, LA, USA, 19–22 April 2010; pp. 1–7.
14. Yurekten, S.; Kara, A.; Mardikyan, K. Energy efficient green transformer manufacturing with amorphous cores. In Proceedings of the International Conference on Renewable Energy Research and Applications (ICRERA), Madrid, Spain, 20–23 October 2013; pp. 534–536.
15. De Lucia, M.; Messeri, M. Method and Device for Manufacturing Transformers with a Core Made of Amorphous Material, and Transformer thus Produced. WO Patent 2016/142504, 15 September 2016.
16. IEEE Standards Association. *IEEE Guide for Loss Evaluation of Distribution and Power Transformers and Reactors, IEEE Std C57.120-2017 (Revision of IEEE Std C57.120-1991)*; IEEE: New York, NY, USA, 2017; pp. 1–53.
17. Winston Official Site, Prismatic LFP Cell. Available online: http://en.winston-battery.com/index.php/products/power-battery/category/lithium-ion-power-battery (accessed on 27 July 2020).
18. Toshiba Official Site, High Power Prismatic LTO Cell. Available online: https://www.scib.jp/en/product/cell.htm (accessed on 27 July 2020).
19. Simcenter Magnet Software Official Site. Available online: https://www.plm.automation.siemens.com/global/it/products/simcenter/magnet.html (accessed on 27 July 2020).

20. Kabitz, S.; Gerschler, J.B.; Ecker, M.; Yurdagel, Y.; Emmermacher, B.; Andrè, D.; Mitsch, T.; Sauer, D.U. Cycle and calendar life study of a graphite $LiNi_{1/3}Mn_{1/3}Co_{1/3}O_2$ Li-ion high energy system. Part A: Full cell characterization. *J. Power Source* **2013**, *239*, 572–583. [CrossRef]

21. Ecker, M.; Nieto, N.; Kabitz, S.; Schmalstieg, J.; Blanke, H.; Warnecke, A.; Sauer, D.U. Calendar and cycle life study of Li(NiMnCo)$O_2$-based 18650 lithium-ion batteries. *J. Power Source* **2014**, *248*, 839–851. [CrossRef]

22. Omar, N.; Monem, M.A.; Firouz, Y.; Salminen, J.; Smekens, J.; Hegazy, O.; Gaulous, H.; Mulder, G.; Van den Bossche, P.; Coosemans, T.; et al. Lithium iron phosphate based battery—Assessment of the aging. *Appl. Energy* **2014**, *113*, 1575–1585. [CrossRef]

23. Antonelli, M.; Ceraolo, M.; Desideri, U.; Lutzemberger, G.; Sani, L. Hybridisation of rubber tired gantry (RTG) cranes. *J. Energy Storage* **2017**, *12*, 186–195. [CrossRef]

24. Fritzson, P. *Introduction to Modeling and Simulation of Technical and Physical Systems with Modelica*, 1st ed.; Wiley-IEEE Press: New York, NY, USA, 2011; ISBN 9781118094259.

25. Ceraolo, M.; Funaioli, M.; Lutzemberger, G.; Pasquali, M.; Poli, D.; Sani, L. Electrical Storage for the Enhancement of Energy and Cost Efficiency of Urban Railroad Systems. In Proceedings of the Second International Conference on Railway Technology: Research, Development and Maintenance, Ajaccio, France, 8–11 April 2014; pp. 1–17.

*Article*

# Optimal Siting and Sizing of Wayside Energy Storage Systems in a D.C. Railway Line

**Regina Lamedica [1], Alessandro Ruvio [1,*], Laura Palagi [2] and Nicola Mortelliti [1]**

[1]   Department of Astronautics, Electrical and Energetic Engineering, Sapienza University of Rome, 00184 Rome, Italy; regina.lamedica@uniroma1.it (R.L.); mortelliti.1831574@studenti.uniroma1.it (N.M.)
[2]   Department of Computer, Control, and Management Engineering, Sapienza University of Rome, 00185 Rome, Italy; laura.palagi@uniroma1.it
*   Correspondence: alessandro.ruvio@uniroma1.it

Received: 22 October 2020; Accepted: 25 November 2020; Published: 27 November 2020

**Abstract:** The paper proposes an optimal siting and sizing methodology to design an energy storage system (ESS) for railway lines. The scope is to maximize the economic benefits. The problem of the optimal siting and sizing of an ESS is addressed and solved by a software developed by the authors using the particle swarm algorithm, whose objective function is based on the net present value (NPV). The railway line, using a standard working day timetable, has been simulated in order to estimate the power flow between the trains finding the siting and sizing of electrical substations and storage systems suitable for the railway network. Numerical simulations have been performed to test the methodology by assuming a new-generation of high-performance trains on a 3 kV direct current (d.c.) railway line. The solution found represents the best choice from an economic point of view and which allows less energy to be taken from the primary network.

**Keywords:** optimization; energy storage system (ESS); siting; sizing; regenerative braking; particle swarm optimization (PSO) algorithm; net present value (NPV); railway network

---

## 1. Introduction

The need to respect and safeguard the world has led issues such as sustainability to become the main focus in order to reduce the environmental impact generated by human activities [1]. In particular, in the railway sector, researchers are focusing on the development of new solutions and techniques to improve the efficiency and systems' capacity to achieve higher energy savings [2]. One of the possible strategies is the recovery of the energy produced by the trains during the braking phases [3]. However, a coordination between traction and braking phases is needed. Contrariwise the regenerated energy must be dissipated in the rheostats, which leads to a significant loss of energy efficiency.

For this reason, it is necessary to increase the receptivity of the system, that is its ability to accept braking energy, by installing:

- Reversible substations (RSSs): allow bidirectional power flows (from the primary network to the contact line and vice versa), so the regenerated energy can be sent to the network to be used by the operator (Figure 1a [4]).
- Energy storage systems (ESSs): store braking energy surplus and return it to the contact line when necessary (Figure 1b [4]), thus limiting the energy taken from the network [5].

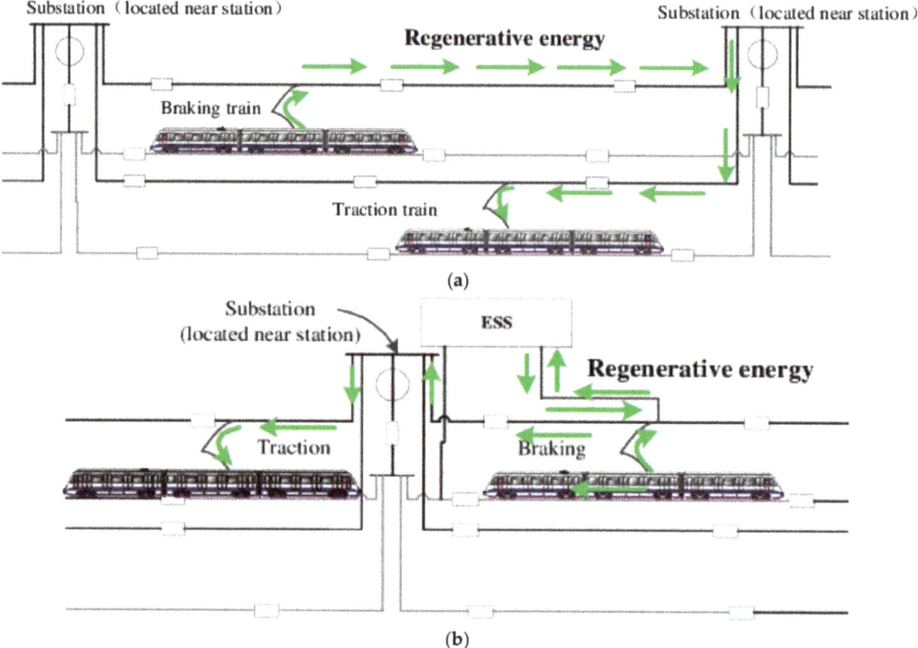



**Figure 1.** Recovery of braking energy: (**a**) reversible substations (RSSs); (**b**) energy storage systems (ESSs).

More in detail, there are two types of energy storage systems:

- Mobile ESSs, located on board vehicles.
- Wayside ESSs, installed in electrical substations or along the track, which don't place weight or size constraints and which balance the voltage in the weak points of the network [6].

Energy storage systems can bring benefits from an energy and environmental point of view (less energy taken from the primary grid, lower consumption of fossil fuels). The problem of sizing an ESS has been addressed under many aspects. First of all, from the point of view of the location: installation solutions have been provided both on board the train and along the railway line [7–15]. The wayside ESS installations provide the advantage of not overloading the trains and their sizing has been analyzed from different points of view:

- Only technical, with optimization algorithms to minimize the energy taken from the primary network but without carrying out an economic analysis of the investment [16–21].
- Technical-economic, with solutions that minimized, for example, the annual cost of electricity or installation cost [22–24].

The second is more complete and adherent to reality; an oversized ESS can work considering the energy saving, but it makes the investment not convenient if the system never provides a positive economic return.

The evaluation of investments in financial mathematics through two methodologies, namely the net present value (NPV) and the payback period (PBP) could be very attractive.

The purpose of the paper is to use the NPV as the objective function of an optimization algorithm, to solve an engineering problem. The correct sizing and positioning of an ESS through an economic approach. The solution provided finds the maximum NPV to obtain the greatest economic returns, at this point the PBP in which the investment is repaid is calculated, as a further economic indicator of the advantage of the installation.

This solution, the best from an economic point of view, should then be approached from a technical one, to see what appreciable innovations it produces from this point of view.

Through the use of a railway simulation software it is possible to calculate the energy taken from the network and the energy losses in the system, thus comparing these parameters between the solution found and the initial case (without ESS installed) it is also possible to evaluate the technical benefits of the proposed solution.

Optimal siting and sizing of ESSs are investigated deeply in literature [7,8,16–34].

A short dissertation of literature references compared with the proposed approach is presented.

In [7] a genetic algorithm (GA) is used to find the optimal sizing of a storage system placed on board the train minimizes the energy withdrawn from the network, maximizing energy saving. From an economic point of view the method shown an investment very onerous. It also highlights long simulation times with the GA, while the choice of the PSO in our paper is aimed precisely at reducing this drawback.

Reference [11] finds the optimal sizing of a storage system located along the line through a genetic algorithm (GA) that minimizes the energy taken from the grid. In this case several solutions are offered between obtaining greater energy savings or lower economic costs, but in any case the energy withdrawn from the network is less than [7]. The methodology proposed in our paper shows that the ESS position could be found near one of the several electrical substations.

In [21] a particle swarm optimization (PSO) algorithm find the best position and size of a supercapacitor minimizing energy taken from the grid. The study does not carry out an economic analysis of the investment and the model allows to use only supercapacitors. The approach proposed in our paper allows to foresee the use of different types of storage for the ESS to be installed.

In [23] the PSO algorithm is used with the aim of minimizing the annual cost of energy, obtaining as output the optimal sizing and position of the ESS. The paper has a techno-economic approach similar to that proposed by our paper, with the use of the PSO algorithm to minimize the objective function; however, it uses a parameter that is not part of the investment evaluation methodologies of financial mathematics, unlike the NPV and PBP proposed by our paper.

In [30] an optimization algorithm has been used to find the optimal sizing of an ESS minimizing its installation cost but not considering the best position. Moreover, the costs are based on load time series and not using a railway simulator, as in our case.

There are also studies in which the PSO is not the best choice, for example in the reference [31] it is highlighted that the PSO has a robustness, calculated in terms of efficiency, of 65% in the calculation of the position and optimal sizing of a hybrid PV system with storage, therefore it was decided to use crow search optimization (CSO) algorithm which instead shows an efficiency of 90% in this study.

In [33] the PSO shows an increase in costs and in convergence times (when the uncertainty on the generation capacity from renewable sources rises) higher than in the general algebraic modeling system (GAMS). for the correct positioning and sizing of ESS and capacitor bank in a microgrid Therefore, the second method is preferred.

By comparing different optimization algorithms for positioning and dimensioning of ESS in a radial distribution network, in [34] it is highlighted how the PSO has a shorter execution time than the others but the solution found is not the best. For this reason, teacher learning-based optimization (TLBO) is defined as more satisfactory for the case in question.

The paper proposes a new method to determine the optimal configuration (position and size) of one or more ESSs along the route, with the aim of maximizing the return on investment. The numerical simulations performed on an extra-urban railway system, with the aid of two simulation tools using a realistic traffic model, show the effectiveness of the proposed method. The resulting problem is a nonlinear integer optimization model where some of the involved functions are a black box, namely are known only by means of the use of a simulation toolbox which does not allow the use of standard mixed integer nonlinear programming (MINLP) methods [35] that make use of derivatives.

For this reason, the optimal configuration is carried out with a heuristic optimization algorithm, the particle swarm optimization (PSO), which has proved to be very flexible and successful in dealing with computationally intensive and highly nonlinear problems, such as the one presented by railway simulation. In [36] two optimization algorithms, dynamic programming optimization (DPO) and particle swarm optimization (PSO), to minimize the energy taken from the grid in a metropolitan network are compared, finding that the second takes about 60% less time to find the solution. The simulation time is a choice factor in the use of the PSO algorithm also for our proposed method.

*Novelty of Approach Proposed*

The proposed approach aims to jointly determine both the optimal size of the energy storage systems and their position along a railway line, maximizing the economic benefits.

A technical-economic model is included into a software developed by the Authors, to establish the economic convenience of the found solution. The main novelty of the proposed approach is to use an optimization algorithm not to minimize the energy purchased from the network, but to maximize the economic benefits from an ESS's installation in a railway line. The methodology has been designed to be performed with any type of d.c. electrified railway line and train, using a software that provides the dynamic and cinematic characteristic of the train and load flow calculation for electrical analysis. The procedure has been implemented in MATLAB™ software (MATLAB 2019b (v9.7), The MathWorks, Inc., Natick, MA, USA).

The structure of the paper is as follows: Section 2 presents the simulation software and models used, the optimization problem is formulated and the algorithm used to solve it is explained. Section 3 presents the case study and shows the results obtained. In Section 4 the obtained results are discussed and compared with the others that are presented in the literature, furthermore future research directions are proposed. Finally, in Section 5 presents our conclusions.

## 2. Methods

### 2.1. The Train Performance Simulation

Train performance is calculated in an Excel™ (Excel 2019 (v16.0), Microsoft Corporation, Redmond, WA, USA) environment [37–39]; the software follows the basic laws of motion. It requires:

- The plan-altimetry characteristics of the line, i.e. the slope of the line with respect to a zero-horizontal line (expressed in ‰), in addition to the radius of curvature of the curves and the length of the tunnels on the analyzed route.
- The electro-mechanical characteristics of rolling stock, in order to simulate various types of trains along the line.

From these inputs, the software provides dynamic and kinematic profiles of the individual vehicle as outputs. It schematizes the vehicle as a material point that absorbs a power at a given point of the line. From an energy point of view, turnouts or level crossings are interpreted by the simulation software as points at which it is necessary to change direction. Therefore is necessary simulate another line and then interconnect it. Information entry related to the operating mode of the line determines the reference traffic scenario and the program provides the system time graph as output.

Setting manually the headway, the code processes the time of departure on each route for each vehicle. On the basis of the set interval for processing, the pre-set traffic scenario is simulated. Starting from the calculation of braking energy, obtainable for every stop, the software computes the theoretically recoverable energy for each train.

Table 1 shows a summary of the main parameters calculated taken into account input data.

**Table 1.** Inputs and outputs of train performance simulation.

| Characteristics of the Route (Inputs) | Characteristics of the Vehicle (Inputs) | Values Every 15 Seconds (Outputs) |
|---|---|---|
| Progressive<br>Slopes<br>Curves<br>Tunnels<br>Speed limits | Aerodynamic resistance<br><br>Traction force<br>Braking force<br>Weight | Pantograph power<br><br>Distance |

### 2.2. The Electrical Model of the Traction System

The electrical computation software allows to solve load flow calculations for a d.c. electrified network. The software for each simulation step, based on kilometer points of vehicles, automatically creates an equivalent electric network in which the nodes represent the electrical substations, vehicles and parallel points; the branches are the traction equivalent circuit stretches between the above-mentioned nodes.

The following models are adopted for the network's computation, even in presence of braking energy recovery. The models and the procedure is well presented in [37].

- Electric substation: Figure 2 shows the V-I characteristic and the equivalent circuit adopted for a typical Substation. When the Substation is disconnected (zone 2), at node A is imposed the $P = 0$ condition. In supply or regeneration status, the node is set with the $V = V_0$ condition (no-load output voltage) or $V = V_{rec}$ (network recovery voltage) and the equivalent resistance $R_2$ or $R_1$ is introduced, respectively, for the different voltage drop in the inverter and rectifier bridges.
- Vehicles in traction: the node is normally set with the $P$ = constant condition. If the pantograph voltage decreases and the current exceeds the maximum value allowed by the electric drive, the iterative computation to resolve the non-linear equations system is repeated imposing the constraint $I = I_{max}$, i.e. constant current absorption. The model follows the V-P characteristic shown in Figure 3. $V_{min}$ is the minimum line voltage allowed to contain voltage drops while $V_{max}$ is the maximum line voltage allowed for which the system can receive the braking power supplied by the rolling stock.
- Braking vehicles: if the system is receptive, all the power is fed into the traction line, as long as the node voltage, resulting from the solution of the system of equations, allows it. In fact, if the $V_{max}$ value is reached, only a portion of the recoverable power is fed into the line. This voltage is a non-linear function of the voltage in the nodes and in the network conductances. The difference between the recoverable power and that actually recovered represents the power dissipated on the braking resistors.

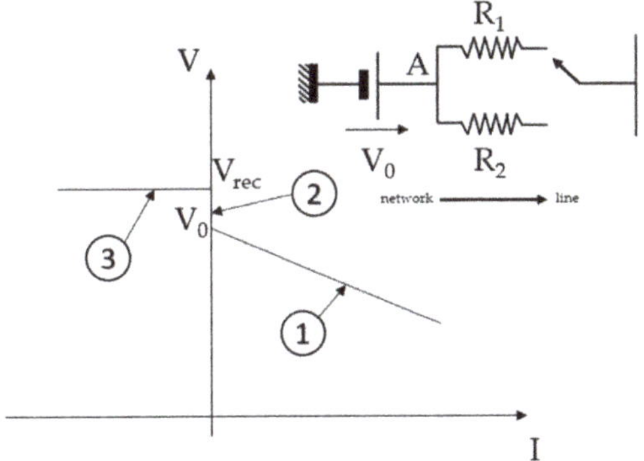


**Figure 2.** V-I characteristic of a typical electric substation.

The software can also simulate the behavior of a mixed rheostatic-regenerative braking (dash-dot line in Figure 3). Like the model shows, with this type of braking as the pantograph voltage increases, the current dissipated on the rheostats increases; at the same time, the current fed into the traction line decreases until it is erased once the maximum voltage value ($V_{max}$) is reached.

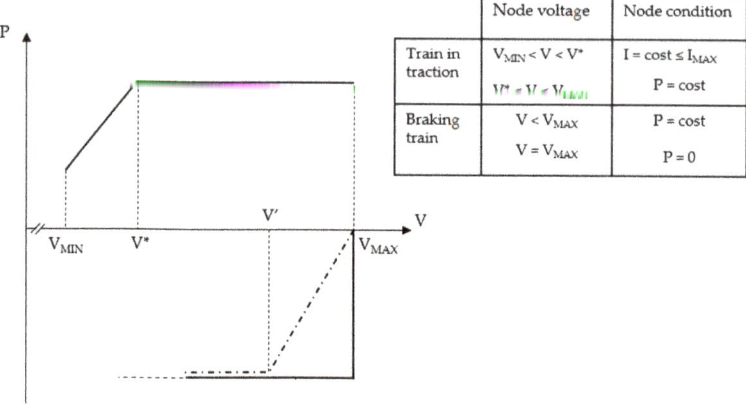

|  | Node voltage | Node condition |
|---|---|---|
| Train in traction | $V_{MIN} < V < V^*$ <br> $V^* < V < V_{MAX}$ | $I = \text{cost} \leq I_{MAX}$ <br> $P = \text{cost}$ |
| Braking train | $V < V_{MAX}$ <br> $V = V_{MAX}$ | $P = \text{cost}$ <br> $P = 0$ |

**Figure 3.** V-P characteristic of traction vehicles.

n nodes network are expressed directly by a set of values of the n independent variables, i.e., of the voltages or currents, or voltages and currents (at the different nodes), the remaining n dependent variables (currents, voltages) can be obtained respectively with the following equations:

$$I_j = \sum_{j=1}^{n} G_{ji} V_i \tag{1}$$

$$V_i = \sum_{j=1}^{N} R_{ij} I_j \tag{2}$$

where $R_{ij}$ and $G_{ji}$ are the coefficients of the matrices of the node resistances and conductances, respectively.

$G_{ji}$, relating to a generic $I_j$, is obtained by setting equal to zero the voltages of all the nodes except the i-th node, and will therefore be:

$$G_{ji} = \frac{I_j}{V_i} \quad (V_1 = V_2 = \ldots = V_{i-1} = V_{i+1} = \ldots = V_n = 0) \tag{3}$$

Similarly for $R_{ij}$, setting $I_j \neq 0$:

$$R_{ij} = \frac{V_i}{I_j} \left(I_1 = I_2 = \ldots = I_{j-1} = I_{j+1} = \ldots = I_n = 0\right) \tag{4}$$

The linear equations written in the form:

$$x_i = f_i\left(x_1 \ldots x_j \ldots x_n\right) \tag{5}$$

i.e., each one explicit with respect to one of the variables x, are solved by using the Gauss-Seidel method: the simulation software is thus able to calculate the line voltage ($V_{LINE}$).

Using the same approach, the software determines the State of Charge of an ESS ($SoC_{ESS}$) during the simulation through the SoC update formula shown in Equation (6):

$$SoC_{ESS}(t) = SoC_{ESS}\,(t=0) + \frac{1}{3600 \cdot Energ_{acc}} \int_0^t V_{ESS}(\tau) \cdot I_{ESS}(\tau) d\tau \tag{6}$$

where $V_{ESS}$ and $I_{ESS}$ are the voltage and current of the ESS, respectively, calculated by the simulation tool, while $Energ_{acc}$ is the nominal energy of the ESS.

The output of the software provides the output values of powers from each electrical substation, the maximum recoverable powers during vehicle braking, the energy storage operation profile and also the line voltage profile and the currents flowing in the substation feeders. Table 2 summarizes inputs and outputs of the simulation tool.

**Table 2.** Inputs and outputs of traction system simulation.

| Electrical Inputs | Outputs |
|---|---|
| Number and kilometric point of substations | Instantaneous powers required to substations |
| Number and type of transformer-rectifier units | Voltage at the vehicle pantographs, at substation bars and at parallel points |
| Catenary feature | Line minimum instantaneous voltages |
| Line rated, minimum and maximum voltage | Current on substation feeders and in traction circuit branches |
| Traffic Input | Recovered energy on line |
| Train performance traffic file | Power losses in traction circuit |
| | Recovered energy in the network |
| | Energy storage operating profile |

### 2.3. Economic Model

The methodology developed uses the net present value (NPV) of a storage system as a target function. The net present value (NPV) of an investment at time $n = 0$ (today) is equal to the discounted cash flow ($C_n$) from year $n = 1$ to $n = N$ (end of useful life of the accumulator) minus the amount of the investment ($I_0$) at the start of the investment ($n = 0$) and the replacement costs in the useful life period ($C_{rep}$). The NPV is mathematically expressed as:

$$NPV = -I_0 + \sum_{n=1}^{N} \frac{C_n}{(1 + ir)^n} - C_{rep} \tag{7}$$

The cash flow $C_n$ of year "n" is equal to all the income $I_n$ minus the operating and maintenance expenses $E_n$ of that year:

$$C_n = I_n - E_n \tag{8}$$

$I_n$ and $E_n$ are expressed by the following equations:

$$I_n = \left(Energ_{noacc\_n} - Energ_{withacc\_n}\right) \cdot C_{ch} \tag{9}$$

$$Energ_{noacc\_n} = Energ_{noacc\_day} \cdot N_d \tag{10}$$

$$Energ_{noacc\_day} = \left(Energ_{noacc\_peak} \cdot N_{pp}\right) + \left(Energ_{noacc\_soft} \cdot N_{sp}\right) \tag{11}$$

$$Energ_{withacc\_n} = Energ_{withacc\_day} \cdot N_d \tag{12}$$

$$Energ_{withacc\_day} = \left(Energ_{withacc\_peak} \cdot N_{pp}\right) + \left(Energ_{withacc\_soft} \cdot N_{sp}\right) \tag{13}$$

$$E_n = C_f \cdot P_{acc} + C_v \cdot Energ_{day}^{acc} \cdot N_d + \frac{Energ_{day}^{acc}}{n_{CH}} C_{CH} \tag{14}$$

The first term represents the fixed costs of the ESS according to the installed power; the second term the variable costs according to the energy supplied by the ESS in a year. The third term represents the recharge costs of the ESS, a function of the energy delivered in a working day, the cost coefficient of electricity and the average charging efficiency of the ESS:

$$Energ_{day}^{acc} = (Energ_{peak}^{acc} \cdot N_{pp}) + (Energ_{soft}^{acc} \cdot N_{sp}) \tag{15}$$

$Energ^{acc}_{peak}$ and $Energ^{acc}_{soft}$, as well as $Energ_{noacc\_peak}$, $Energ_{noacc\_soft}$, $Energ_{withacc\_peak}$ and $Energ_{withacc\_soft}$, are taken from the railway simulator when simulating a railway line with an ESS, of which values of position, nominal power and nominal energy are inserted.

When the NPV found by the algorithm is a positive solution, the payback period (PBP), which allows to calculate the time within the capital invested ($I_0$) in the purchase of a medium-long-cycle production factor is recovered through the net financial flows generated ($C_n$) is also calculated to obtain a further indicator of the economic convenience of the investment:

$$PBP = \frac{I_0}{C_n} \tag{16}$$

$I_0$ is the capital cost of the accumulator and is:

$$I_0 = C_P \cdot P_{acc} + C_E \cdot Energ_{acc} \tag{17}$$

$C_{rep}$ is the future value of replacement cost and is shown by Equation (18) [40]:

$$C_{rep} = C_{RP}[(1 + ir)^{-L_R} + (1 + ir)^{-2L_R} + \ldots + (1 + ir)^{-num \cdot L_R}] \tag{18}$$

where "num" is the number of times the battery must be replaced during the life of the ESS.

The software is able to simulate, in a standard working day, the charge and discharge cycles followed by the ESS. It is possible to obtain the duration ($n_{batt}$) of the batteries; once the useful life (N) of the ESS. Using Equation (19) it is possible calculate how many times it is necessary to replace the batteries:

$$num = \frac{N}{n_{batt}} \tag{19}$$

*2.4. Optimization*

2.4.1. Problem formulation

The NPV of the system, which has been explained in more detail in Section 2.3, is considered as the objective function: the variable $x_c$ on which the NPV depends is the set consisting of the nominal value of the power $P_{acc}$ and energy $Energ_{acc}$ of the ESS, as well as the position Pos where the ESS itself is installed along the track:

$$x_c = \{P_{acc}, Energ_{acc}, Pos\} \tag{20}$$

The study is focused on finding the solution that has the maximum NPV, so the optimization problem is formulated as follows:

$$Z = \max_{x_c} NPV(x_c) \tag{21}$$

subject to the following constraints:

$$P_{acc}^{min} \leq P_{acc} \leq P_{acc}^{max} \tag{22}$$

$$\text{Energ}_{\text{acc}}^{\min} \leq \text{Energ}_{\text{acc}} \leq \text{Energ}_{\text{acc}}^{\max} \tag{23}$$

$$0 \leq \text{Pos} \leq \text{L}^{\max} \tag{24}$$

$$\text{V}_{\text{LINE}}^{\min} \leq \text{V}_{\text{LINE}} \leq \text{V}_{\text{LINE}}^{\max} \tag{25}$$

$$\text{I}_{\text{ESS}}^{\min} \leq \text{I}_{\text{ESS}} \leq \text{I}_{\text{ESS}}^{\max} \tag{26}$$

$$\text{V}_{\text{ESS}}^{\min} \leq \text{V}_{\text{ESS}} \leq \text{V}_{\text{ESS}}^{\max} \tag{27}$$

$$\text{SoC}_{\text{ESS}}^{\min} \leq \text{SoC}_{\text{ESS}} \leq \text{SoC}_{\text{ESS}}^{\max} \tag{28}$$

where $\text{L}_{\max}$ is the track length, $\text{V}_{\text{LINE}}{}^{\min}$ and $\text{V}_{\text{LINE}}{}^{\max}$ are the minimum and maximum allowable line voltage values, $\text{I}_{\text{ESS}}$, $\text{V}_{\text{ESS}}$ and $\text{SoC}_{\text{ESS}}$ are the current, voltage, and SoC of to the ESS, respectively calculated using the software described in Section 2.2 [37–39]; each of them is limited to its minimum and maximum value.

The optimization of the siting and sizing of the ESS is formulated as a non-linear integer problem (MINLP) whose non-linear objective function is evaluated by means of a simulation tool, which is able to verify the safe operating conditions of the d.c. feeder system, and it is not known in analytic form. Thus it fits in the class of black-box non convex MINLP problems which are among the most challenging optimization models. The solution of black box problems requires the use of derivative-free algorithms that do not require the derivatives not even in approximate form. Further non convexity of the functions makes it difficult to use exact method for integer programs.

The great computational difficulty and the computational time needed in evaluating the objective function suggest using simple heuristic algorithms to explore the solution space quickly. Hence the optimization problem is addressed using a Particle Swarm Optimization-based algorithm as the solution method.

### 2.4.2. PSO Algorithm

Particle swarm optimization (PSO) is a heuristic algorithm inspired by the choreography of a flock of birds [41]. In the space of real numbers, every single possible solution is modeled as a particle moving through the hyperspace of the problem. At each iteration, the velocities of individual particles are stochastically adjusted based on the best historical position for the particle itself and the best position in the neighborhood. Both the best particle and the neighborhood best one are found based on a user defined fitness function. The flow chart of the integration of the simulation software with the proposed PSO algorithm for the positioning and dimensioning is shown in Figure 4. The characteristics of the route and those of the rolling stock (expressed in Table 1) are provided as input to the software described in Section 2.1. The power supplied by the supply line is then estimated by calculating the consumption of the vehicle. The traffic file produced is the input to the electrical calculation software described in Section 2.2 (together with the other values listed in Table 2 which are entered by the user) who is able to calculate the energy supplied by the substations for a particular configuration of location and sizing. The objective function in Equation (21) is thus calculated and finally, the PSO returns a feasible and better solution than the original one.

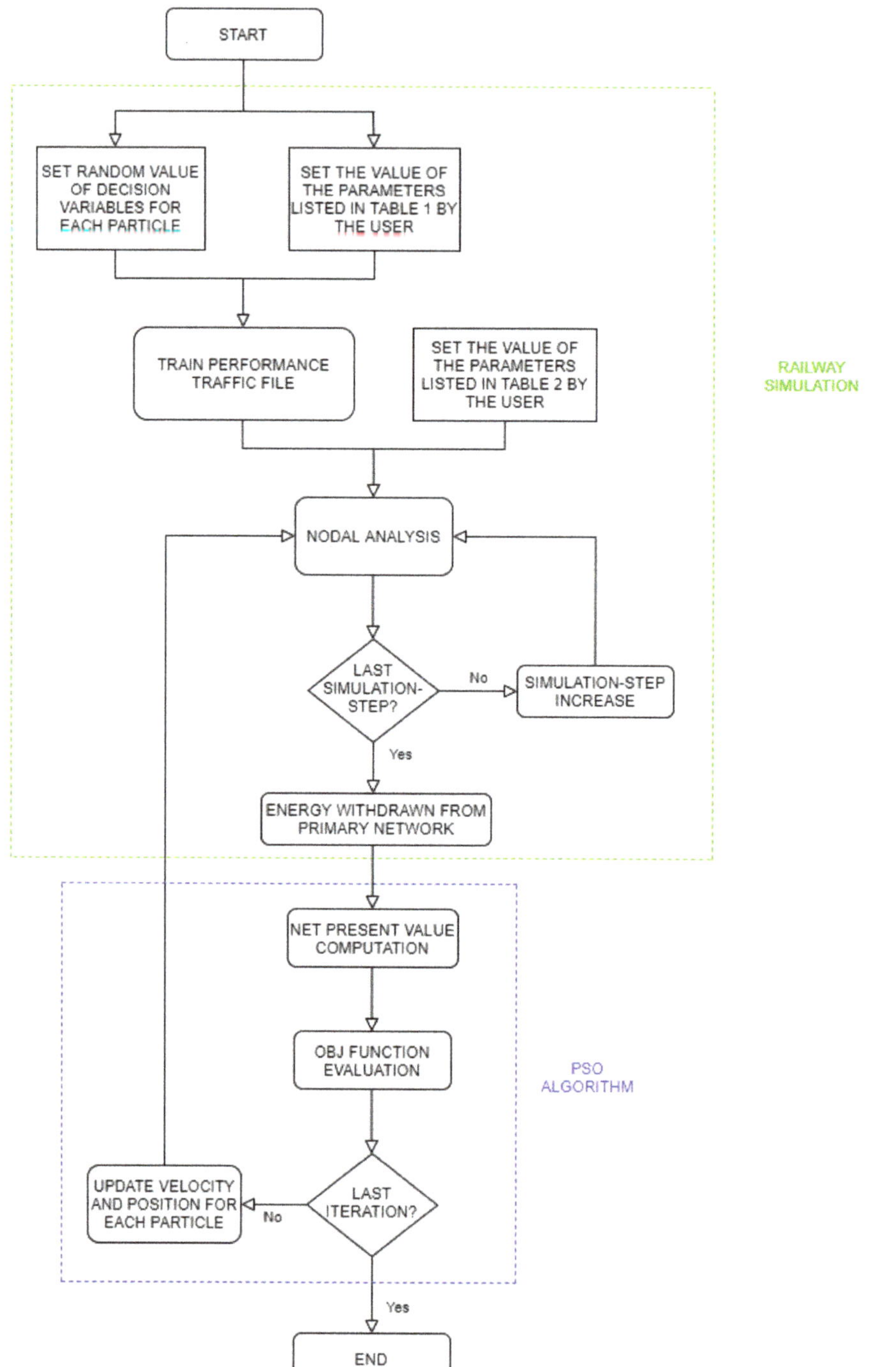



**Figure 4.** Flowchart of the PSO-based solution algorithm.

## 3. Results

### 3.1. Case Study

In order to establish the effectiveness of the proposed solution method, a case study calculated on an extra-urban railway line is presented. The NPV of the system is evaluated in 3 cases:

- Without possibility of recovering the braking energy by trains running on the railway line and without ESSs along the line.
- With the possibility of recovering the braking energy by trains and without ESS along the line.
- With the possibility of recovering the braking energy and with ESS installed along the line.

The PSO algorithm is coded in MATLAB™ (MATLAB 2019b (v9.7), The MathWorks, Inc., Natick, MA, USA) in the "Global Optimization Toolbox" library and was run by entering the following parameters:

- Fun: objective function, specified as a function handle or function name (in this case the NPV calculation explained in Section 2.3).
- Nvars: number of variables, specified as a positive integer (for the proposed model it is equal to 3). The solver passes row vectors of length "nvars" to "fun".
- Lb: lower bounds, specified as a real vector or array of doubles.
- Ub: upper bounds, specified as a real vector or array of doubles.
- Options: options for "particleswarm" function, in particular the swarm size and the maximum number of iterations.

### 3.1.1. Topological Characteristics of Railway Line

Figure 5 shows the line: on the x-axis there is the kilometric progressive, while on the y-axis there is the height above sea level. The green line shows the planimetric profile of the railway line, which has a length of 75.1 km. Railway stations are highlighted with red circles while the electric substations along the route are highlighted with blue circles.

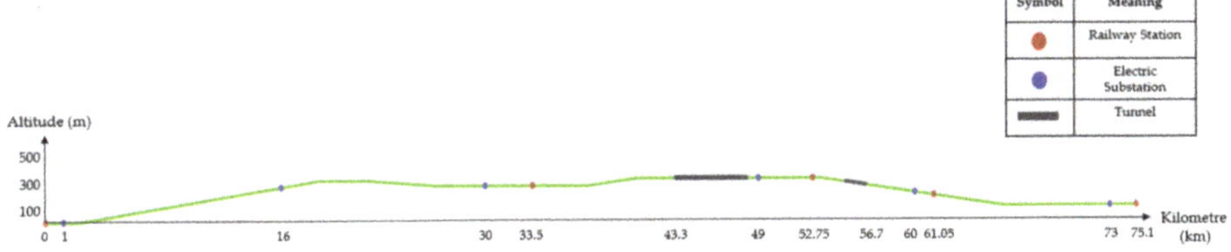


**Figure 5.** Characteristics of the railway line.

Between the arrival and departure stations there are three intermediate stations, as shown in the Table 3. As can be seen from Figure 5, there are two tunnels on the line, highlighted with gray rectangles, whose characteristics are shown in Table 4.

**Table 3.** Position of the stations.

| Station Name | Progressive (km) |
| --- | --- |
| Departure station | 0 |
| Intermediate station #1 | 33.5 |
| Intermediate station #2 | 52.75 |
| Intermediate station #3 | 61.05 |
| Arrival station | 75.1 |

**Table 4.** Tunnels' position.

| Tunnel Name | Length (km) | From (km) | To (km) |
| --- | --- | --- | --- |
| Tunnel #1 | 5 | 43.3 | 48.3 |
| Tunnel #2 | 1.5 | 55.2 | 56.7 |

Finally, Figure 6 shows the slopes of the line.

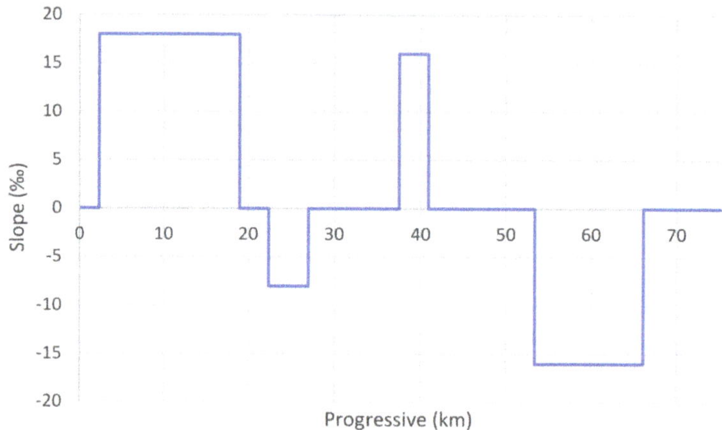

**Figure 6.** Slope of line.

### 3.1.2. Electrical Characteristics

The line is 3 kV d.c. electrified, according to CEI EN 50163. The contact line consists of two copper contact wires, each with a diameter of 11.8 mm and a section of 150 mm². They are supported by two carrying ropes in copper stranded with 19 wires each with a diameter of 2.8 mm and a section of 120 mm² and is standardized according to CEI EN 50119, CEI EN 50122-1, CEI EN 50122-2, and CEI EN 50149. Each substation, regulated according to CEI EN 50388, is equipped with two conversion units of 5.4 MW and a value of 0.13 Ω per unit has been set for the internal resistance. The characteristics and location of the substations are shown in the Table 5.

**Table 5.** Substations on the line.

| Substation Name | Conversion Units | Nominal Power (MW) | Progressive (km) |
| --- | --- | --- | --- |
| Electric substation #1 | 2 | 5.4 | 1 |
| Electric substation #2 | 2 | 5.4 | 16 |
| Electric substation #3 | 2 | 5.4 | 30 |
| Electric substation #4 | 2 | 5.4 | 45 |
| Electric substation #5 | 2 | 5.4 | 60 |
| Electric substation #6 | 2 | 5.4 | 73 |

### 3.1.3. Rolling Stock Characteristics

A high-performance train with characteristics similar to the ETR 1000 model of the "Ferrovie dello Stato Italiane" Group, whose parameters are shown in Table 6, moves in the two opposite directions following the same speed cycle (Figure 7). The speed limits imposed on the train are highlighted with a black line. Blue line highlights the speed profile of the train.

**Table 6.** Rolling stock data.

| Parameter | Value | |
| --- | --- | --- |
| Loaded weight | 500 | t |
| Rotating mass coefficient | 0.05 | |
| Motor efficiency | 0.85 | |
| Auxiliaries power | 400 | kW |
| Full speed | 360 | km/h |
| Maximum traction effort | 370 | kN |
| Maximum deceleration | 0.6 | m/s$^2$ |

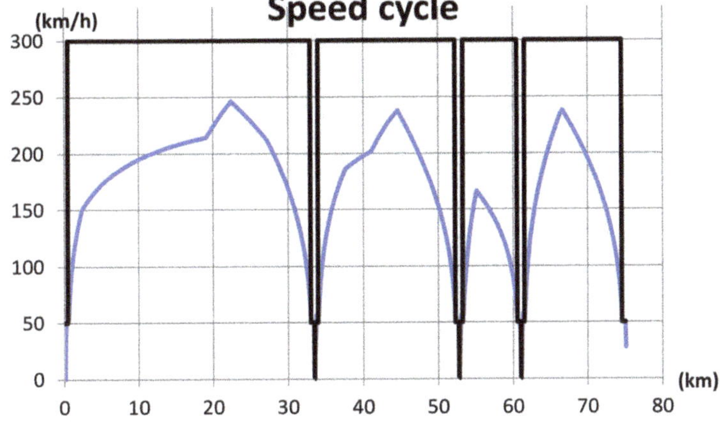


**Figure 7.** Speed cycle of the train.

### 3.1.4. Timetable

A standard working day has two bands, respectively "peak" and "soft", divided as follows:

- During the "peak" period, lasting 2 hours, every 7 minutes a train departs from the departure and arrival stations.
- During the "soft" period, also lasting 2 hours, every 30 minutes a train departs from the departure and arrival stations.

A standard working day has been divided into eight periods (three peak and five soft), as shown in Table 7, according to the greatest demand that occurs at the beginning and end of the day.

**Table 7.** Service hours.

| From (hh:mm) | To (hh:mm) | Band |
| --- | --- | --- |
| 07:00 | 09:00 | Peak |
| 09:00 | 11:00 | Soft |
| 11:00 | 13:00 | Soft |
| 13:00 | 15:00 | Peak |
| 15:00 | 17:00 | Soft |
| 17:00 | 19:00 | Peak |
| 19:00 | 21:00 | Soft |
| 21:00 | 23:00 | Soft |

### 3.2. Numerical Results

Several simulations are carried out, imposing different upper and lower bounds of the input variables, for the case study described in Section 3.1 to highlight the strengths and weaknesses of the methodology implemented into the software. 20,000 iterations and 25 particles are imposed by running the proposed algorithm on a device with an Intel®Core ™ i7 processor (2.20 GHz, 64 bit), 16 GB of RAM and MATLAB™ R2019b. The economic inputs of ESS based on Li-ion battery, detailed in [23], are shown in Table 8.

**Table 8.** Economic inputs.

| Parameters | Values | Unit |
|---|---|---|
| $C_p$ | 120 | €/kW |
| $C_E$ | 440 | €/kWh |
| $C_f$ | 7.2 | €/kW |
| $C_v$ | 0.0011 | €/kWh |
| $C_{ch}$ | 0.015 | €/kWh |
| N | 20 | #years |
| $\eta_{CH}$ | 0.90 | - |

For the cost coefficient of electricity ($C_{ch}$) a flat rate of 0.015 €/kWh has been considered. It is a realistic value compared to the tariff regime used in Italy for railways.

The variables managed by the program are: position, nominal power and nominal energy; the position value is chosen from those included between the beginning and the end of the line, the others are parameterized as shown in Table 9.

**Table 9.** Input parameters of the optimization algorithm.

| Variables of Decision | Lower Bound | Upper Bound |
|---|---|---|
| Position (km) | 0 | 75.1 |
| Nominal power (MW) | 0.1 | 1 |
| Nominal energy (MWh) | 0.1 | 1 |

The solution found by the optimization:

- NPV: 391,577.59 €
- Payback period: 2.844 years
- ESS's position: 55 km
- Nominal power: 0.400 MW
- Nominal energy: 0.200 MWh

The solution provided allows to have the maximum possible profit and a relatively short payback period. Furthermore, making a comparison between the case without ESS and with ESS, less energy is drawn from the primary network and also a lower impact on the electrical substations, which are, especially those located in the middle of the line, less stressed, as shown in Figure 8b. Specifically, being able to count on the possibility of trains to reuse braking energy decreases the energy absorbed by the primary network by about 34% and the installation of an ESS along the line allows for about a further 8% less energy drawn. The electrical substations, on the other hand, thanks to the recovery of braking energy, deliver on average 30% less power (with peaks of over 40% for the substations located at the limits of the railway line), which is further lowered by about 5% in average (with peaks over 10% for the mid-line substations) thanks to the presence of ESS.

Moreover, it should be noted that through the ESS the electrical substations and the traction line are stressed less, with consequent benefits from the point of view of maintenance, aging and replacement of materials.

Even if it is not one of the specific objectives of the model, the adoption of an ESS along the line has the further benefit of reducing the system's energy losses (31 MWh against 32.9 MWh in a standard working day).

These aspects allow considerable savings in the long period and increase the advantages that can be obtained after installation: these are not taken into account by energy savings, but are estimated in the economic model on which PSO is based, so the solution provided allows to have greater overall benefits than looking only at the energy aspect.

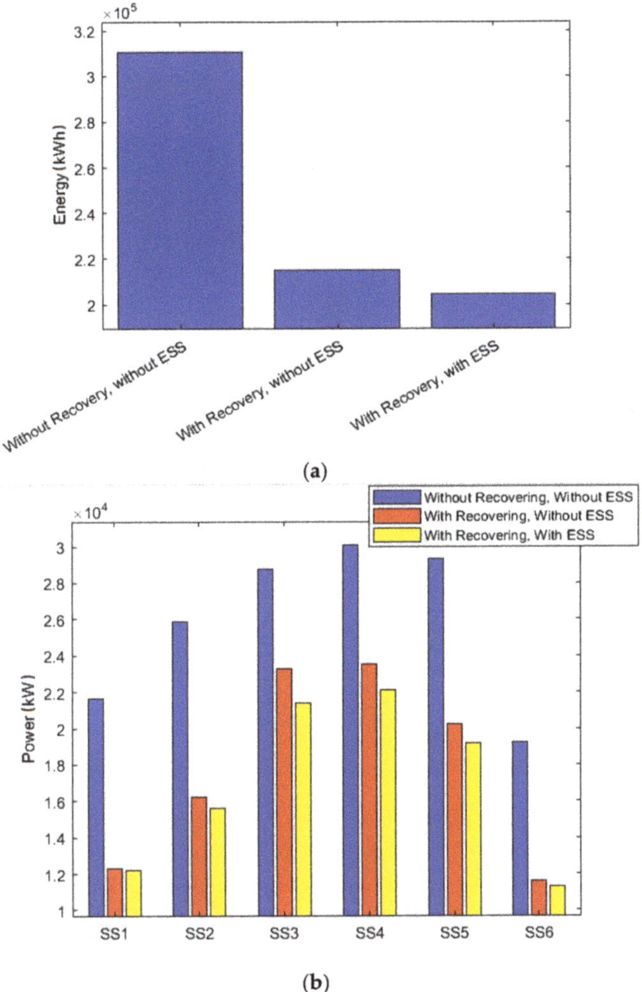


**Figure 8.** Graphic software output: (**a**) Total energy withdrawn from the primary network in a working day; (**b**) Power supplied by each substation present on the line.

The operation of the line (timetable) is a fundamental role in the choice of the installation to be carried out. Using the same network and route characteristics of previous simulation, but with another timetable (only "soft" periods), it possible to appreciate that a much higher percentage of energy saving, taken into account ESS, is reached (Figure 9).

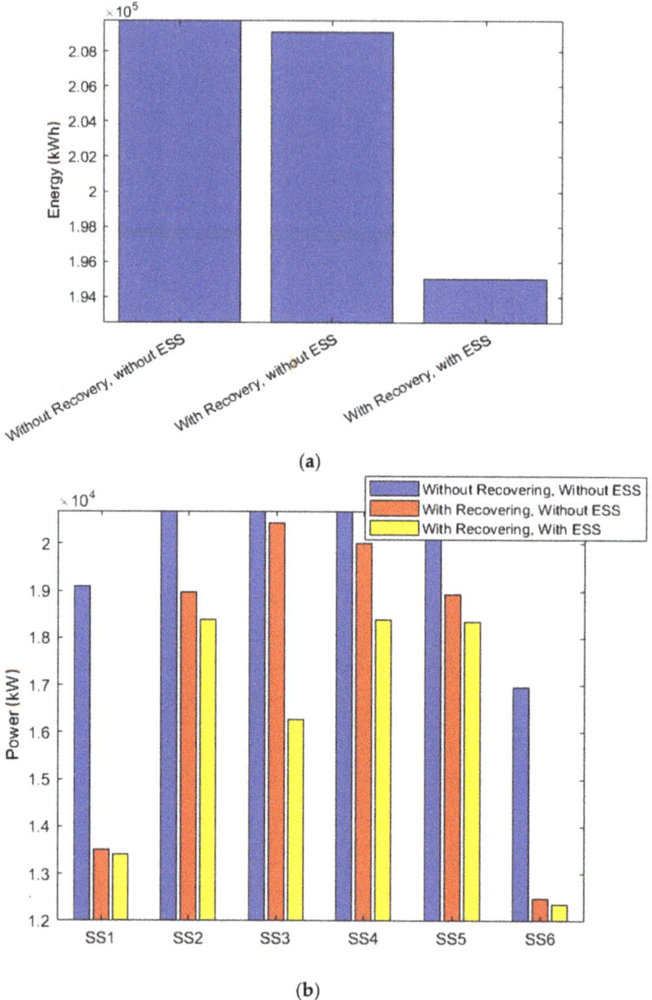



**Figure 9.** Graphic software output: (**a**) Total energy withdrawn from the primary network in a working day; (**b**) Power supplied by each substation present on the line.

The electrical substations load is more decreased, with benefits from the point of view of maintenance and replacement, as previously mentioned, which are not directly visible in energy consumption. Being able to simulate multiple service hours, it can be seen how this characteristic of the proposed model allows for a more accurate and realistic solution.

To avoid unusable solutions and excessively long and expensive simulation times, it is advisable to choose numerical ranges that have real ESS sizes and that are not too wide.

A further strength of the software is the possibility of providing more ESSs on the line: in fact, it is possible to impose a number of accumulators greater than 1 and to obtain the optimal solution from the software with that specific configuration. A simulation performed by predicting 2 ESSs in line, with the same ranges of decision variables predicted in Table 8, led to the follow solution:

- NPV: 383,451.71 €
- Payback period: 3.6648 years

- ESSs' position: 6 km, 55 km
- Nominal power: 0.100 MW, 0.400 MW
- Nominal energy: 0.100 MWh, 0.200 MWh

The proposed solution involves a lower NPV than the case with only one ESS installed, but looking at the graphic outputs of the software (Figure 10), comparing the case without ESSs with that with two ESSs, an energy saving of about 14% is noted (compared to 8% in the previous case). The electrical substations deliver about 9% less power on average (before it was 5%), with peaks over 15% for the mid-line substations when 2 ESSs are installed according to the proposed method.

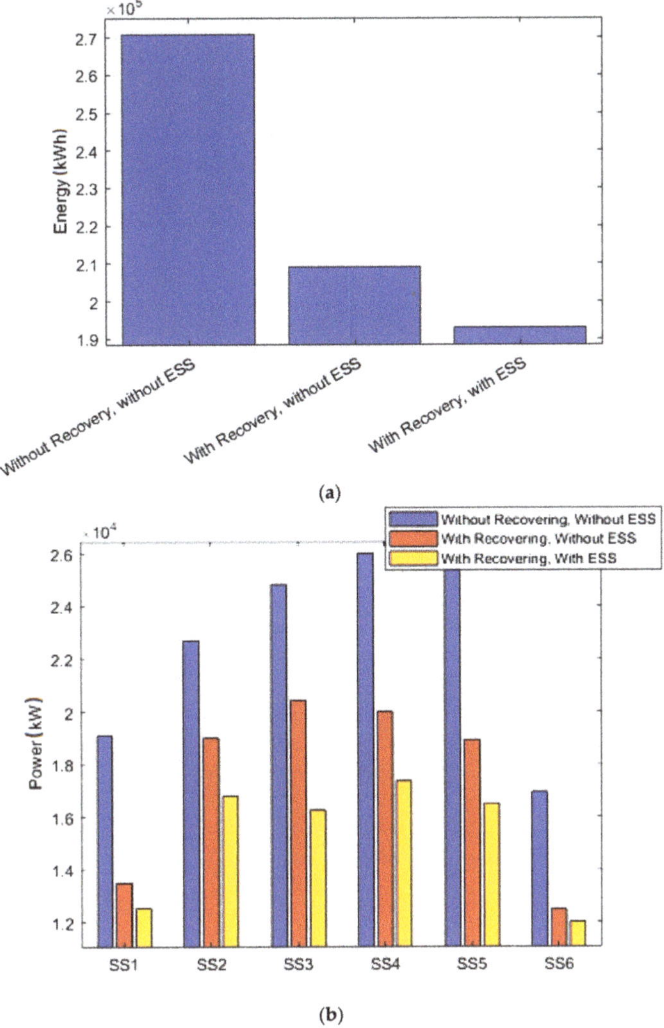


**Figure 10.** Graphic software output with 2 ESSs on the route: (**a**) Total energy withdrawn from the primary network in a working day; (**b**) Power supplied by each substation present on the line.

Also, from the point of view of the energy losses of the system, the solution proposed with two ESSs installed results in lower losses compared to the previous case with only one ESS installed (29.3 MWh against 31 MWh in a standard working day).

Despite a lower economic return from the installation, i.e. a slightly lower but still positive NPV, it is possible to lighten the electrical substations and draw even less energy from the network. The user who manages the software can evaluate the benefits of this solution through the graphic output, that allows to have all the information, both technical and economic, of the installation to be carried out.

## 4. Discussion

The results obtained show how the proposed solution allows an effective economic return on investment. From an economic point of view, also as regards the energy one, it allows savings percentages comparable to those obtained from [7] and [25], but with the advantage of having stationary installations that don't burden the convoys. The PBP turns out to be greater than that found by our model since in [25] the economic aspect of the problem is a sub-objective, not the main one and therefore is not maximized.

Compared to other research based on single types of accumulation [16,20,21,23] by varying the parameters characterizing the different types within the model it is possible to consider different ESSs, and the different investments can be compared with each other by means of a single value, which is the NPV.

The use of this parameter also allows, as seen, to evaluate cases with different numbers of ESS installed, which entail different energy savings (shown to the user) but in any case, comparable from an economic point of view through the NPV. Comparing different scenarios is a possibility also offered by [30] but it needs real traffic data: the use of the railway simulator instead allows to simulate countless traffic scenarios only by knowing the characteristics of the line and those of the rolling stock. The model takes into account the management of the line and offers different solutions based on its use.

The economic model described in Section 2.3 in the evaluation of the NPV examines the charge and discharge cycles of the ESS as [24], which bases its optimization on this parameter. The solution found by our model is therefore based on a greater number of variables and this allows for a more in-depth and realistic solution.

Further developments in the sector could involve an ever-greater integration between technical and economic design, using multi-objective optimization to simultaneously maximize or minimize technical aspects (e.g. energy losses or system receptivity) and economic aspects (e.g. the capital cost of the investment or the expected economic returns). Modified and more complex models should be adopted to effectively describe the reciprocal influence between the two aspects mentioned above.

From the point of view of railway safety, many studies [42–49] have recently been conducted on the safety of the infrastructure and on the effects that accidents and derailments have on it and on its deterioration. The addition of ESSs on the line involves new variables whose effects should be deepened.

Furthermore, it would be interesting to be able to compare other solutions, such as on-board storage systems or the installation of reversible electrical substations, with those found. Common models should be developed for the different types of proposed solutions in which future technological developments can also be taken into account.

The proposed methodology can be further developed by providing for the inclusion in the model of constraints related to the exchange of energy or by inserting a limit on the energy consumed in order to have lower energy consumption and less stress on the electrical substations. Other solutions to explore include: the replacement of the PSO with derivative-free algorithms of "pattern search" type, based on the guided exploration of the feasible region; instead of the railway simulator, use non-parametric machine learning models derived from the data collected or simulated on the network. In the last case, it would apply algorithms which use the derivatives and that are more advanced and efficient for the determination of the optimal solution or one of its good approximation.

Since the installation of an ESS involves a very expensive investment, the initial choice of the type of storage is an important problem: in fact, the rates of the economic parameters change according to the technology used and this means that, for the same installed size, different technologies involve different costs and therefore also different economic benefits. The proposed methodology allows to take into account the impact of the various rates, i.e., the various technologies that can be installed, since the economic model used by the algorithm to calculate the NPV of the system depends on economic parameters that change according to the technology adopted, therefore by changing the parameters' values of ESS it is possible to obtain a different economic calculation and a different optimal solution, depending on the chosen technology.

## 5. Conclusions

This paper proposes a methodology to determine the optimal siting and sizing of energy storage systems along railway lines: the results found show how the proposed approach maximizes the economic return on investment required to install an ESS along a railway line. Using the PSO it is possible to obtain a faster solution than other meta-heuristic algorithms. Using the NPV and the PBP, used by the analysis of investments in financial mathematics, it is possible to directly compare two alternative investments with the same risk profile, thus obtaining a more realistic assessment of the feasibility and convenience of the installation.

The proposed method can be used for any type of d.c. electrified railway line, it can model any existing rolling stock and allows to simulate multiple working timetables and the simultaneous presence of multiple ESSs. In this way it is possible to find the optimal equipment for countless ESSs' installations located along a railway line through a model that is also easy to use.

**Author Contributions:** Conceptualization, R.L., L.P. and A.R.; methodology, N.M.; software, N.M.; validation, L.P.; formal analysis, N.M.; investigation, N.M.; data curation, N.M. and A.R.; writing—original draft preparation, N.M.; writing—review and editing, R.L. and A.R.; visualization, N.M.; supervision, R.L. and A.R. All authors have read and agreed to the published version of the manuscript.

**Funding:** This research received no external funding.

**Conflicts of Interest:** The authors declare no conflict of interest.

## Abbreviations

| | |
|---|---|
| $I_0$ | Capital cost of the ESS (€) |
| $n$ | Generic year during the life of the ESS (-) |
| $N$ | Useful life (in years) of the ESS (-) |
| $C_n$ | Cash flow of the year "n" (€) |
| $I_n$ | Income of the year "n" (€) |
| $E_n$ | Expenses of the year "n" (€) |
| $N_d$ | Total days of use of the ESS within one year (-) |
| $N_{pp}$ | Number of peak periods per day (-) |
| $N_{sp}$ | Number of soft periods per day (-) |
| $Energ_{noacc\_n}$ | Total energy withdrawn from the primary network, without ESSs in the track, in the year "n" (kWh) |
| $Energ_{withacc\_n}$ | Total energy withdrawn from the primary network, with ESSs in the track, in the year "n" (kWh) |
| $Energ_{noacc\_day}$ | Total energy withdrawn from the primary network, without ESSs in the track, in one working day (kWh) |
| $Energ_{withacc\_day}$ | Total energy withdrawn from the primary network, without ESSs in the track, in one working day (kWh) |

| | |
|---|---|
| $\text{Energ}_{\text{noacc\_peak}}$ | Total energy withdrawn from the primary network during the peak period, without ESSs in the track (kWh) |
| $\text{Energ}_{\text{noacc\_soft}}$ | Total energy withdrawn from the primary network during the soft period, without ESSs in the track (kWh) |
| $\text{Energ}_{\text{withacc\_peak}}$ | Total energy withdrawn from the primary network during the peak period, with ESSs in the track (kWh) |
| $\text{Energ}_{\text{withacc\_soft}}$ | Total energy withdrawn from the primary network during the soft period, with ESSs in the track (kWh) |
| $C_{\text{ch}}$ | Cost of electricity coefficient (€/kWh) |
| $C_{\text{f}}$ | Fixed costs of the ESS (€/kW) |
| PBP | Payback Period (in years) of the ESS (-) |
| $P_{\text{acc}}$ | Nominal power of the ESS (MW) |
| $C_{\text{v}}$ | Variable costs of the ESS (€/kWh) |
| $\text{Energ}_{\text{day}}^{\text{acc}}$ | Total energy supplied by the ESS in a working day (kWh) |
| $\text{Energ}_{\text{peak}}^{\text{acc}}$ | Energy supplied by the ESS during the peak period (kWh) |
| $\text{Energ}_{\text{soft}}^{\text{acc}}$ | Energy supplied by the ESS during the soft period (kWh) |
| $n_{\text{CH}}$ | Average value of the battery charging efficiency (-) |
| $C_{\text{P}}$ | Specific cost of battery power (€/kW) |
| $C_{\text{E}}$ | Specific cost of battery energy (€/kWh) |
| $\text{Energ}_{\text{acc}}$ | Nominal energy of the ESS (MWh) |
| $C_{\text{rep}}$ | Future value of replacement cost (€) |
| $C_{\text{RP}}$ | Future value of the battery (assumed to be equal to the capital cost of the battery) (€) |
| ir | Interest rate (-) |
| $L_{\text{R}}$ | Battery replacement period (in years) (-) |
| $n_{\text{batt}}$ | Useful life (in years) of the batteries (-) |
| NPV | Net present value |
| num | Number of times the battery must be replaced during the life of the ESS (-) |

## References


1. European Comission. EU Energy, Transport and GHG Emissions Trends to 2050, Reference Scenario 2016. Available online: http://ec.europa.eu/energy (accessed on 26 November 2020).
2. Aguado, J.A.; Sánchez Racero, A.J.; de la Torre, S. Optimal Operation of Electric Railways with Renewable Energy and Electric Storage Systems. *IEEE Trans. Smart Grid* **2018**, *9*, 993–1001. [CrossRef]
3. Lamedica, R.; Ruvio, A.; Papalini, A.; Spalvieri, C.; Rossetta, I. Train braking impact on energy recovery: The case of the 3 kV d.c. railway line Roma-Napoli via Formia. In Proceedings of the 2019 AEIT International Annual Conference, Florence, Italy, 18–20 September 2019; pp. 1–6.
4. Su, S.; Tang, T.; Wang, Y. Control of urban rail transit equipped with ground-based supercapacitor for energy saving and reduction of power peak demand. *Energies* **2016**, *9*, 1–19.
5. Lamedica, R.; Ruvio, A.; Galdi, V.; Graber, G.; Sforza, P.; Guidi Buffarini, G.; Spalvieri, C. Application of Battery Auxiliary Substations in 3kV Railway Systems. In Proceedings of the 2015 AEIT International Annual Conference, Naples, Italy, 14–16 October 2015; pp. 1–6.
6. Gao, Z.; Fang, J.; Zhang, Y.; Jiang, L.; Sun, D.; Guo, W. Evaluation of strategies to reducing traction energy consumption of metro systems using an Optimal Train Control Simulation model. *Int. J. Elec. Power* **2015**, *67*, 439–447. [CrossRef]
7. Kampeerawat, W.; Toseki, T. Integrated design of smart train scheduling, use of onboard energy storage, and traction power management for energy-saving urban railway operation. *IEEJ J. Ind. Appl.* **2019**, *8*, 893–903. [CrossRef]
8. Wu, C.; Lu, S.; Xue, F.; Jiang, L.; Chen, M. Optimal Sizing of Onboard Energy Storage Devices for Electrified Railway Systems. *IEEE Trans. Transport. Electrific.* **2020**, *6*, 1301–1311. [CrossRef]
9. Popescu, M.; Bitoleanu, A. A Review of the Energy Efficiency Improvement in DC Railway Systems. *Energies* **2019**, *12*, 1092. [CrossRef]

10. Meinert, M. New mobile energy storage system for rolling stock. In Proceedings of the 2009 European Conference on Power Electronics and Applications, Barcelona, Spain, 8–10 September 2009; pp. 1–10.
11. Kampeerawat, W.; Toseki, T. Efficient urban railway design integrating train scheduling, wayside energy storage, and traction power management. *IEEJ J. Ind. Appl.* **2019**, *8*, 915–925. [CrossRef]
12. Barsali, S.; Bechini, A.; Giglioli, R.; Poli, D. Storage in electrified transport systems. In Proceedings of the 2012 IEEE International Energy Conference and Exhibition, Florence, Italy, 9–12 September 2012; pp. 1003–1008.
13. Hanmin, L.; Gildong, K.; Changmu, L.; Euijin, J. Field tests of DC 1500 V stationary energy storage system. *Int. J. Railway* **2012**, *5*, 124–128.
14. Ratniyomchai, T.; Kulworawanichpong, T. A Demonstration Project for Installation of Battery Energy Storage System in Mass Rapid Transit. *Energy Procedia* **2017**, *138*, 93–98. [CrossRef]
15. Barrero, R.; Tackoen, X.; Van Mierlo, J. Stationary or onboard energy storage systems for energy consumption reduction in a metro network. *Proc. Inst. Mech. Eng. F J. Rail Rapid Transit* **2010**, *224*, 207–225. [CrossRef]
16. Gee, A.M.; Dunn, R.W. Analysis of trackside flywheel energy storage in light rail systems. *IEEE Trans. Veh. Technol.* **2015**, *64*, 3858–3869. [CrossRef]
17. Lee, H.; Song, J.; Jang, G.; Kim, G.; Lee, H.; Lee, C. Capacity optimization of the supercapacitor energy storages on DC railway system using a railway powerflow algorithm. *Int. J. Innov. Comput. I.* **2011**, *7*, 2739–2753.
18. Ratniyomchai, T.; Hillmansen, S.; Tricoli, P. Optimal capacity and positioning of stationary supercapacitors for light rail vehicle systems. In Proceedings of the 2014 International Symposium on Power Electronics, Electrical Drives, Automation and Motion, Ischia, Italy, 18–20 June 2014; pp. 807–812.
19. Battistelli, L.; Ciccarelli, F.; Lauria, D.; Proto, D. Optimal design of DC electrified railway stationary storage system. In Proceedings of the 2009 International Conference on Clean Electrical Power, Capri, Italy, 9–11 June 2009; pp. 739–745.
20. Ratniyomchai, T.; Hillmansen, S.; Tricoli, P. Energy loss minimisation by optimal design of stationary supercapacitors for light railways. In Proceedings of the 2015 International Conference on Clean Electrical Power, Taormina, Italy, 16–18 June 2015; pp. 511–517.
21. Calderaro, V.; Galdi, V.; Graber, G.; Piccolo, A. Optimal siting and sizing of stationary supercapacitors in a metro network using PSO. In Proceedings of the 2015 IEEE International Conference on Industrial Technology, Seville, Spain, 17–19 March 2015; pp. 1–6.
22. Xia, H.; Chen, H.; Yang, Z.; Lin, F.; Wang, B. Optimal Energy Management, Location and Size for Stationary Energy Storage System in a Metro Line Based on Genetic Algorithm. *Energies* **2015**, *8*, 11618–11640. [CrossRef]
23. Graber, G.; Calderaro, V.; Galdi, V.; Piccolo, A.; Lamedica, R.; Ruvio, A. Techno-economic Sizing of Auxiliary-Battery-Based Substations in DC Railway Systems. *IEEE T. Transp. Electr.* **2018**, *4*, 616–625. [CrossRef]
24. De la Torre, S.; Sanchez-Racero, A.J.; Aguado, J.A.; Reyes, M.; Martianez, O. Optimal Sizing of Energy Storage for Regenerative Braking in Electric Railway Systems. *IEEE Trans. Power Syst.* **2015**, *30*, 1492–1500. [CrossRef]
25. Wang, B.; Yang, Z.; Lin, F.; Zhao, W. An improved genetic algorithm for optimal stationary energy storage system locating and sizing. *Energies* **2014**, *7*, 6434–6458. [CrossRef]
26. Li, Y.; Wang, J.; Gu, C.; Liu, J.; Li, Z. Investment optimization of grid-scale energy storage for supporting different wind power utilization levels. *J. Mod. Power Syst. Clean Energy* **2019**, *7*, 1721–1734. [CrossRef]
27. Yuan, Z.; Wang, W.; Wang, H.; Yildizbasi, A. A new methodology for optimal location and sizing of battery energy storage system in distribution networks for loss reduction. *J. Energy Storage* **2020**, *29*, 101368. [CrossRef]
28. Miao, D.; Hossain, S. Improved gray wolf optimization algorithm for solving placement and sizing of electrical energy storage system in micro-grids. *ISA Trans.* **2020**, *102*, 376–387. [CrossRef]
29. Kerdphol, T.; Fuji, K.; Mitani, Y.; Watanabe, M.; Qudaih, Y. Optimization of a battery energy storage system using particle swarm optimization for stand-alone microgrids. *Int. J. Elec. Power.* **2016**, *81*, 32–39. [CrossRef]
30. Yoshida, Y.; Figueroa, H.P.; Dougal, R.A. Use of Time Series Load Data to Size Energy Storage Systems. In Proceedings of the 2018 IEEE Green Energy and Smart Systems Conference, Long Beach, CA, USA, 29–30 October 2018; pp. 1–6.
31. Pandey, A.K.; Kirmani, S. Optimal location and sizing of hybrid system by analytical crow search optimization algorithm. *Int. Trans. Electr. Energy Syst.* **2020**, *30*, e12327. [CrossRef]

32. Ruiyang, J.; Jie, S.; Jie, L.; Wei, L.; Chao, L. Location and Capacity Optimization of Distributed Energy Storage System in Peak-Shaving. *Energies* **2020**, *13*, 513.
33. Rajamand, S. Loss cost reduction and power quality improvement with applying robust optimization algorithm for optimum energy storage system placement and capacitor bank allocation. *Int. J. Energy Res.* **2020**, *44*, 11973–11984. [CrossRef]
34. Lata, P.; Vadhera, S. TLBO-based approach to optimally place and sizing of energy storage system for reliability enhancement of radial distribution system. *Int. Trans. Electr. Energ. Syst.* **2020**, *30*, e12334. [CrossRef]
35. Davis, E. Solving MINLP containing noisy variables and black-box functions using Branch-and-Bound. *Comput. Aided Chem. Eng.* **2006**, *21*, 633–638.
36. Calderaro, V.; Galdi, V.; Graber, G.; Piccolo, A. Deterministic vs. Heuristic Algorithms for Eco-Driving Application in Metro Network. In Proceedings of the 2015 International Conference on Electrical Systems for Aircraft, Railway, Ship Propulsion and Road Vehicles, Aachen, Germany, 3–5 March 2015; pp. 1–6.
37. Capasso, A.; Lamedica, R.; Ruvio, A.; Giannini, G. Eco-friendly urban transport systems. Comparison between energy demands of the trolleybus and tram systems. *Ing. Ferrov.* **2014**, *69*, 329–347.
38. Capasso, A.; Lamedica, R.; Ruvio, A.; Ceraolo, M.; Lutzemberger, G. Modelling and simulation of electric urban transportation systems with energy storage. In Proceedings of the 2016 IEEE 16th International Conference on Environment and Electrical Engineering, Florence, Italy, 7–10 June 2016; pp. 1–5.
39. Lamedica, R.; Gatta, F.M.; Geri, A.; Sangiovanni, S.; Ruvio, A. A software validation for dc electrified transportation system: A tram line of Rome. In Proceedings of the 2018 IEEE International Conference on Environment and Electrical Engineering and 2018 IEEE Industrial and Commercial Power Systems Europe, Palermo, Italy, 12–15 June 2018; pp. 1–6.
40. Poonpun, P.; Jewell, W.T. Analysis of the cost per kilowatt hour to store electricity. *IEEE Trans. Energy Convers.* **2008**, *23*, 529–534. [CrossRef]
41. Del Valle, Y.; Venayagamoorthy, G.K.; Mohagheghi, S.; Hernandez, J.C.; Harley, R.G. Particle Swarm Optimization: Basic Concepts, Variants and Applications in Power Systems. *IEEE T. Evolut. Comput.* **2008**, *12*, 171–195. [CrossRef]
42. Dindar, S.; Kaewunruen, S.; An, M. Identification of appropriate risk analysis techniques for railway turnout systems. *J. Risk Res.* **2018**, *21*, 974–995. [CrossRef]
43. Alawad, H.; Kaewunruen, S.; An, M. Learning from Accidents: Machine Learning for Safety at Railway Stations. *IEEE Access.* **2020**, *8*, 633–648. [CrossRef]
44. Kaewunruen, S.; Bin Osman, M.H.; Wong, H.C.E. Risk-based maintenance planning for rail fastening systems. *ASCE ASME J. Risk Uncertain. Eng. Syst. A Civ. Eng.* **2019**, *5*, 04019007. [CrossRef]
45. Kampczyk, A. Measurement of the Geometric Center of a Turnout for the Safety of Railway Infrastructure Using MMS and Total Station. *Sensors* **2020**, *20*, 4467. [CrossRef] [PubMed]
46. Wang, H.; Chang, L.; Markine, V.L. Structural Health Monitoring of Railway Transition Zones Using Satellite Radar Data. *Sensors* **2018**, *18*, 413. [CrossRef] [PubMed]
47. Guo, Y.; Zhao, C.; Markine, V.L.; Jing, G.; Zhai, W. Calibration for discrete element modelling of railway ballast: A review. *Transp. Geotech.* **2020**, *23*, 100341. [CrossRef]
48. Dindar, S.; Kaewunruen, S.; An, M. Rail accident analysis using large-scale investigations of train derailments on switches and crossings: Comparing the performances of a novel stochastic mathematical prediction and various assumptions. *Eng. Fail. Anal.* **2019**, *103*, 203–216. [CrossRef]
49. Ye, Y.; Shi, D.; Poveda-Reyes, S.; Hecht, M. Quantification of the influence of rolling stock failures on track deterioration. *J. Zhejiang Univ. Sci. A* **2020**, *21*, 783–798. [CrossRef]

**Publisher's Note:** MDPI stays neutral with regard to jurisdictional claims in published maps and institutional affiliations.

*Article*

# A Preliminary Techno-Economic Comparison between DC Electrification and Trains with On-Board Energy Storage Systems

Regina Lamedica [1], Alessandro Ruvio [1,*], Manuel Tobia [1], Guido Guidi Buffarini [2] and Nicola Carones [2]

[1] Department of Astronautical, Electrical, and Energy Engineering, Sapienza University of Rome, 00185 Rome, Italy; regina.lamedica@uniroma1.it (R.L.); tobiamanuel@gmail.com (M.T.)

[2] Italferr S.p.A., 00155 Rome, Italy; g.guidibuffarini@italferr.it (G.G.B.); n.carones@italferr.it (N.C.)

* Correspondence: alessandro.ruvio@uniroma1.it

Received: 17 October 2020; Accepted: 15 December 2020; Published: 18 December 2020

**Abstract:** The paper presents a preliminary technical-economic comparison between a 3 kV DC railway and the use of trains with on-board storage systems. Numerical simulations have been carried out on a real railway line, which presents an electrified section at 3 kV DC and a non-electrified section, currently covered by diesel-powered trains. Different types of ESS have been analyzed, implementing the models in Matlab/Simulink environment. A preparatory economic investigation has been carried out.

**Keywords:** railway system; energy storage system (ESS); lithium batteries; supercapacitor; Simulink; sizing; catenary-free

## 1. Introduction

The International Energy Agency and the International Union of Railways have stated that the 23.1% of worldwide $CO_2$ emissions from fuel combustion are attributable to the transport sector (8258 million tons of $CO_2$). The breakdown of the emissions in this sector is attributable as follows: 73.2% to road transport, 10.4% to maritime transport, 10.5% to air transport and 3.6% to railways [1]. Regarding the European situation, the amount relating to the transport sector is around 30.4% (71.1% road transport, 13.9% maritime transport, 12.7% air transport and 1.5% railways). In the United States, the transport sector accounts for 34.4% of emissions from fuel combustion, in Japan for 20.6%, in Russia for 17.5%, in India 12.7% and in China for 9.6% [2,3]. According to the Statistical Compendium on Transport published annually by the Commission, net of emissions indirectly attributable to rail transport (which are accounted in electricity generation phase), therefore 44% of total emissions are attributable to road passenger transport, 28% for road freight, while air and maritime transport accounted for 13.0 and 13.4%, respectively. Overall, urban mobility would account for 25% of greenhouse gas emissions from transport, interurban for 54% and intercontinental for 22%) [1,4].

From the point of view of railway line electrification, Europe has 55% of the total length of the railway line electrified. Italy has 72% of electrified lines and 28% with diesel-powered trains [5].

E-mobility will be widespread thanks to the European Union funding programs and capital investments from the automotive industry. However, several technical, financial and social challenges need to be overcome. Although electric vehicles present higher purchase and infrastructure costs, they present lower operating costs than a conventional car [6–8]. Due to the limited autonomy and required charging times, battery electric vehicles are indicated for consumers with a limited daily range, but the technology, especially for storage systems, is constantly improving. In the long term, options like plug-in hybrid vehicles are expected to have a relatively big market share [9].

The worldwide diffusion of electric vehicles is still very limited. One of the main reasons of different trends in the sales of electric cars is the presence in some parts if the world of incentive mechanisms such as purchase incentives and free access to controlled priority circulation zones. Another relevant factor is the presence of proper charging infrastructures, which requires the companies involved to adapt their business models [10].

One of the main target areas of the climate goals is road freight transport, since about 25% of the $CO_2$ emissions from road transport in the European Union are produced by heavy-duty vehicles. Although shifting from road freight transport to electrified railways is a possible solution, several studies indicate that the potential for this is limited. New technologies such as electrified highways (eHighways) are proposed, in which an overhead line supply energy to trucks provided with an integrated pantograph. A two-pole catenary system has been developed, ensuring a stable current transmission at speeds up to 100 km/h [11]. This technology is already being used in several projects in countries such as Sweden and United States [12], and several such projects are in the planning phase.

The design of electrified mass transit systems for urban railway traffic, such as trams, trolleybuses and metros, requires taking into consideration several elements: safety, efficiency, cost and visual impact. Traditional electrified transit systems are based on electrical contact wires, such as active rails or catenary. Catenary-free systems have advantages of as low electromagnetic radiation and good visual design [13] In the absence of fixed traction systems, the rolling stock has on board the energy storage system (e.g., batteries, hydrogen, diesel, ultracapacitors, flywheels). Moreover, catenary-free systems are being consolidated in public transportation, especially to restrict the electrical infrastructure, for example in city historical centers, limiting the environmental impact and reducing costs of electrical installations for traction [14].

In the current state some form of on-board storage system is often used for heavy traction (passenger and freight trains), but generally only to recover braking energy and not to operate in catenary-free mode. Battery-powered locomotives were one of the first technologies to be adopted in the early 1900s but due to the limited storage capacity available at that time, it was decided to focus on electrified solutions. Use thus became limited to mining locomotives and shunting locomotives [15].

Nowadays several studies have focused on the possibility to design new trains to operate autonomously because of the advances in energy storage technologies [16]. Trains with on-board storage systems could be adopted for expansion of already electrified railways, point-to-point connections or commuter transport systems [17–19]. It will be important in the near future to considerate the need for prospective analysis of the marketing of rolling stock approved for bimodal operation (i.e., with and without contact lines supplying different voltages). The diffusion of a new system will strongly depend on the cost and the development of new rolling stock with on board storage technology for traction [20–22].

It is well known that for heavy traction, the electrified solution via catenary represents a consolidated and highly reliable approach that is economically sustainable for high traffic lines [23,24]. Moreover, considering the contribution of renewable sources it is possible to increase the energy savings from the primary network. For this reason, several studies and projects have been carried out highlighting the trend of replacing diesel locomotives with electric locomotives to cover long distances, powered by AC or DC electrification systems [25].

For solutions involving on-board energy storage systems (ESS), several technologies can be used, as shown in Figure 1. Since each one of them presents different characteristics, it is important to choose the most suitable technology for the particular case under study. Currently, there are different types of electrochemical accumulators. Lead-acid accumulators, widely used in the automotive sector, represent an economical and reliable solution. However, they have a very low specific energy. Those with nickel cadmium (Ni-Cd), initially used as substitutes for the aforementioned lead-acid accumulators, have been nowadays abandoned due to the toxicity of the cadmium present in them, and the focus has shifted to nickel-metal hydride accumulators (NiMH), which have the disadvantage of presenting memory effects. Finally, lithium batteries (Li-ion) are characterized by high energy and specific power.

Technologies such as flywheels offer a high specific power and a large number of charge and discharge cycles but they present low specific energy. The use of fuel cells in railway transport systems is currently in the early development state and they present many advantages such as zero emissions fuel and high specific energy [26]. Though fuel cell sources can be coupled with high-power density storage elements [27], the combination of Li-ion batteries with supercapacitors (SC) is currently found to be the most economical solution for ESS [26].

Supercapacitors, also known as ultracapacitors, are an innovative electrical EES that has considerably higher capacity values than those of ordinary capacitors and high specific power. The combination of Li-ion batteries and SC in a hybrid energy storage system (H-ESS) allows one to reduce battery stress and aging effects, depending on the charge and discharge cycles. The sizing of the ESS is strictly linked to the control strategy used for the management of energy resources and is limited by the maximum weight and volume allowed by the rolling stock [28,29].

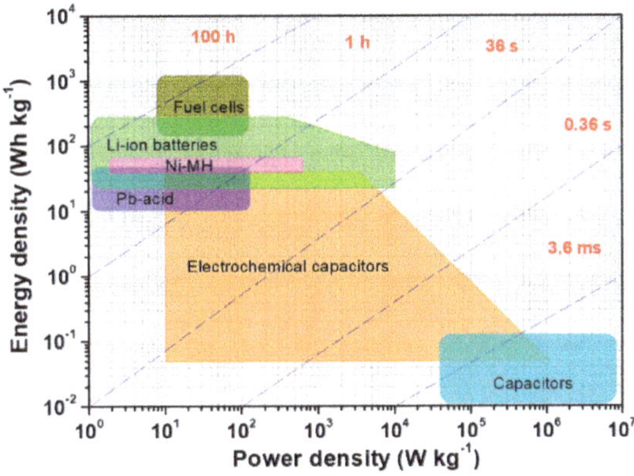


**Figure 1.** Ragone chart [30].

Several studies have been carried out on the applications of storage systems that use lithium-ion cells with high specific power (power-oriented) or high specific energy (energy-oriented). Usually, energy-oriented lithium-ion cells are accompanied by supercapacitors, which deliver the stored energy during power peaks in traction and absorb energy during regenerative braking [31,32]. It is possible to find different applications on electric vehicles [33,34], electric buses [35], small boats [36], water buses [37], and even machines for lifting and displacement of loads [38].

In [39], a battery pack of about 250 kWh and about 2 tons of mass is used on a freight train, in order to solve the problem of the absence of the catenary in the connections of the railway lines to the freight terminals and production plants, avoiding the use of diesel-powered vehicles. In [40], prototypes of battery packs of 200 kWh and about 2 tons are reported for a railway locomotive and of 500 kWh and 5 tons of mass for powering a mining truck. In [41], the authors present a battery pack of 2.22 MWh and 11 tons of mass to power a rolling stock with a total mass of 276 tons that must travel a route of 212 km.

Among the few solutions commercially available for trains equipped with energy storage systems there is a Spanish tramway in Seville, which uses a hybrid storage system consisting of supercapacitors and batteries [42]. In addition, the same train constructor has developed a battery-powered regional train, proposing it as a solution for non-electrified railway lines. The roof-mounted traction batteries provide the power and energy required to propel the train for distances up to 100 km [43,44]. It is important highlight that the adoption of storage system on board trams and trolleybuses is economically convenient for short route sections. There are already design solutions and rolling stock equipped with on-board storage systems available on market.

In addition to the technical and infrastructural characteristics of a given traction system, it is important to carry out a cost assessment, especially when comparing different achievable solutions. In the technical and scientific literature, several studies have been published using the annual cost of energy (ACOE) in order to evaluate the effectiveness in terms of costs of different types of energy resources [45]. In this study, the ACOE is also used to compare the different technical solutions proposed.

This paper presents the evaluation taken into consideration to perform a comparison between direct current electrification and the use of trains with on-board energy storage systems. Numerical simulations have been performed on a real railway line to carry out a preliminary techno-economic comparison showing the models and procedure used. The case study presents an electrified 3 kV DC section and a non-electrified section, currently being covered by diesel-powered trains. For the non-electrified section, the following scenarios have been analyzed: 3 kV DC electrification, since it is the feeding system currently used in the preceding section; high autonomy ESS and ESS with recharge station. For the ESS it has been considered the use of power-oriented Li-ion battery cells and, in the case of hybrid ESS, energy-oriented Li-ion battery cells accompanied by supercapacitors. For the sake of simplicity, only recharging the ESS has been considered, the possibility of swapping the ESS is outside the scope of this work, as it would have an impact on capital costs. It is highlighted that, in cost assessment is not taken into account the cost associated with the purchase of new trains. In this paper, it is assumed that the cost of the on board ESS is the difference cost between traditional electric trains and new trains equipped with on board ESS.

The paper is structured as follows: Section 2 describes the electric models of the vehicle, energy storage systems and DC feeder system, as well as the economic model. Section 3 presents the on-board ESS sizing procedure. Section 4 introduces the simulation software and procedures while Section 5 presents the case study and the numerical results of the simulations. Section 6 concludes the paper.

## 2. Modeling of the Railway System

The proposed model is obtained by using three different sub-models: the railway vehicle and its kinematics, the DC feeder system and the on-board ESS. An economic model is also introduced for calculating the annual energy cost of the proposed solutions.

### 2.1. Train

The longitudinal dynamic of vehicles is evaluated applying Newton's second law and kinematic equations:

$$\begin{cases} m \; \varepsilon \frac{dv}{dt} = F(t) - R_{BASE}(v) - R_{TRACK}(x) \\ x = x_0 + v_0 \, t + \frac{1}{2} \frac{dv}{dt} \, t^2 \\ v = \frac{dx}{dt} \end{cases} \tag{1}$$

where $m$ is the mass of the vehicle, $\varepsilon$ is a correction factor taking into account the rotating mass, $v$ and $x$ are the train speed and position respectively, $F$ is the traction (if positive) or braking (if negative) force. $R_{BASE}(v)$ is the basic resistance including roll resistance and air resistance, and $R_{TRACK}(x)$ is the line resistance caused by track slopes and curves, described by:

$$R_{BASE}(v) = \alpha_1 + \alpha_2 |v| + \alpha_3 v^2 \tag{2}$$

$$R_{TRACK}(x) = mg \, sin(\gamma(x)) + mg \frac{a}{r(x) - b} \tag{3}$$

where $\alpha_1$, $\alpha_2$ and $\alpha_3$ are the coefficients of the Davis formula, related to the train and track characteristics, and they can be estimated by empirical measures; $g$ is the gravitational acceleration and $\gamma(x)$ is the slope grade. The second term of $R_{TRACK}$ is the curve resistance given by empirical formulas, as the Von Röckl's formula, where $r(x)$ is the curvature radius, and $a$, $b$ are coefficients which depend on the track gauge; in this paper it is considered $a = 0.65$ m and $b = 55$ m [46].

From a given a speed cycle, it is possible to calculate the value of the force ($F_{MECH}$) on the wheels required to overcome the vehicle inertia, slopes and curves, aerodynamic friction and rolling friction. Going upstream the vehicle components and their related efficiencies, the power requested to the contact wire $P_{TRAIN}$ is calculated as follows:

$$P_{TRAIN} = \begin{cases} \frac{F_{MECH}\,v}{\eta_t} + P_{AUX}, & F_{MECH} \geq 0 \\ (F_{MECH}\,v)\,\eta_t + P_{AUX}, & F_{MECH} < 0 \end{cases} \tag{4}$$

where $P_{AUX}$ is the power absorbed by board auxiliary services (lighting, cooling or heating), $m$ is the total mass of the train (including the passengers), $v$ is the vehicle speed and $\eta_t$ the total efficiency of the locomotive, which takes into account the efficiency of the gear box, the electric motor and the inverter. To bring into account that the voltage along the track is not constant, the railway vehicle is modeled as an ideal current source $I_{TRAIN}$, whose is calculated as the ratio between vehicle power and line voltage $V_{LINE}$:

$$I_{TRAIN} = \frac{P_{TRAIN}}{V_{LINE}} \tag{5}$$

*2.2. DC Feeder System*

Conventional substations are represented by ideal DC voltage sources, series resistance and series diode if the substations are not reversible [28]. The contact wire is modelled as a set of electric resistances that change their value according to the vehicle position. If $x(t)$ is the train position at the time $t$, the value of the resistance upstream $R_a$ and downstream $R_b$ to the vehicle towards a generic node of the railway feeding system (conventional substation or another train) are calculated by:

$$\begin{cases} R_a(t) = \rho{\cdot}x(t) \\ R_b(t) = \rho{\cdot}[d - x(t)] \end{cases} \tag{6}$$

where $R_a$ and $R_b$ are expressed in [$\Omega$], $\rho$ [$\Omega$/km] represents the resistive coefficient, $d$ [km] is the distance between the two nodes (upstream and downstream the train). In order to improve the train electric model, describing the receptivity of the network under regenerative braking conditions, a small capacitance is connected in parallel to current source that represents the vehicle [46]. In Figure 2 it is shown the electric model of the overall railway system, one side supplied contact line.

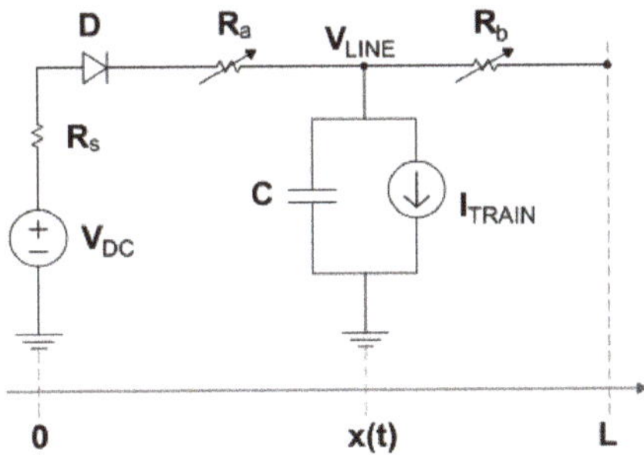


**Figure 2.** Electric model of the one side DC supplied contact line.

*2.3. On-Board ESS*

The electric model of the ESS is reported in Figure 3, it includes the battery and supercapacitor pack, the DC/DC converter and the power flow controller. The on-board ESS is modelled as a pair of ideal current sources describing the battery-based and the SC-based energy storage system, respectively.

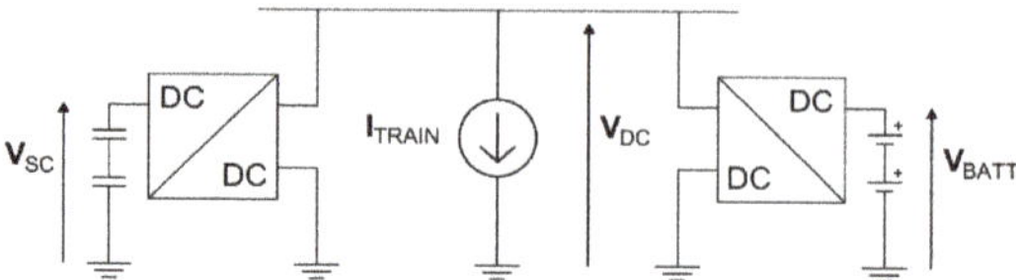

**Figure 3.** On-board Hybrid ESS electric model.

Figure 4a presents the equivalent circuit of the battery pack, which consists of an ideal voltage source that represents the open circuit voltage (OCV), which depends on battery state of charge (SOC); the series resistor $R_{INT}$ represents the internal resistance, whereas $r_d$ and $C_d$ are the RC parallel circuit describing the charge transfer and double layer capacity, respectively. The set of equations that describes the electric model of the battery pack is reported in Equations (7) to (10): the first equation represents Kirchhoff's voltage law, the second one is the n-polynomial relationship between OCV and SOC. The third equation models the SOC update law according to the required current from the battery pack and the last one is the differential equation describing the RC parallel circuit [29].

$$V_{BATT}(t) = OCV(t) - R_{INT}I_{BATT}(t) - u_d(t) \tag{7}$$

$$OCV(SOC) = \beta_n SOC^n + \beta_{n-1} SOC^{n-1} + \beta_0 \tag{8}$$

$$SOC(t) = SOC(t=0) - \frac{1}{3600 \cdot C_{AH}} \int_0^t I_{BATT}(\tau)d\tau \tag{9}$$

$$u_d(t) + r_d C_d \frac{du_d(t)}{dt} = r_d I_{BATT}(t) \tag{10}$$

where $u_d(t)$ is the $r_d C_d$ parallel circuit voltage, $\beta_0 \ldots \beta_n$ are the interpolation coefficients and $C_{AH}$ [Ah] is the battery pack capacity. The electrical model of the SC pack, shown in Figure 4b, consists of the capacitor $C$, modelling SC's capacity; an equivalent series resistance $R_S$ that describes the power loss during the charging and discharging operations; the self-charge resistance $R_L$ models the losses due to the leakage current, which is usually neglected [28,47]:

$$V_{SC}(t) = V_C(t) - R_s I_{SC}(t) \tag{11}$$

$$V_C(t) = V_{SC}(t=0) - \frac{1}{C} \int_0^t I_{SC}(\tau)d\tau \tag{12}$$

$$SOC_{SC}(t) = \frac{1}{3}\left[4 \cdot \left(\frac{V_C(t)}{V_{nom}}\right)^2 - 1\right] \tag{13}$$

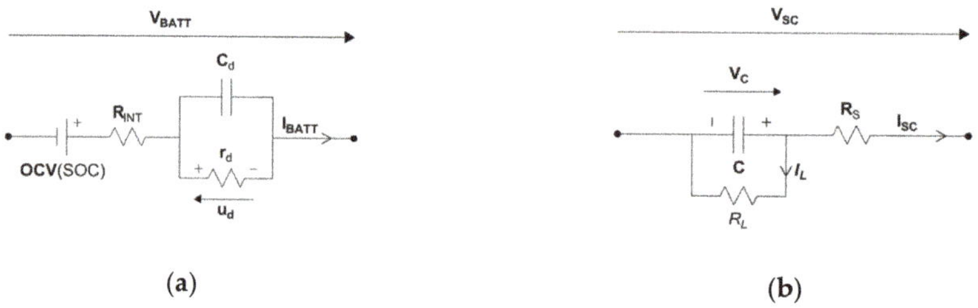

(a)

(b)

**Figure 4.** Equivalent circuit of the battery pack (**a**) and SC pack (**b**).

It is highlighted that, in order to represent the equivalent circuit of a battery or supercapacitor pack, starting from the characteristics associated with the individual cells or modules, it is used an equivalent circuit: $n$ RC blocks in series and $m$ in parallel [48]. The DC/DC converters are modeled by its average efficiency $\eta$ describing power losses. It operates as step-up or step-down converter according to the control characteristic. Given the power reference value $P_{ref}$ provided by the power flow control characteristic, the ESS current value to be delivered during the traction phase is computed by using the following equation:

$$I_{ref} = \frac{1}{\eta}\frac{P_{ref}}{V_t} \tag{14}$$

where $V_t$ is the battery pack or supercapacitor pack voltage.

*2.4. Economic Model*

The annual cost of energy is used to compare the different technical solutions proposed. Therefore, it is necessary to estimate the present value of the total cost, which includes: the cost of capital, the present value of the operating costs and the present value of the replacement cost of the energy storage system [45]. The ACOE is mathematically expressed as:

$$ACOE = CRF{\cdot}C_{TOT} \tag{15}$$

where $CRF$ is the capital recovery factor converting a present value into a stream of equal annual payments over a specified lifetime $N$, at a specified interest rate $r$, and $C_{TOT}$ is the present value of the total cost. The capital recovery factor is computed by using the following equation:

$$CRF = \frac{r(1+r)^N}{(1+r)^N - 1} \tag{16}$$

2.4.1. Costs of an Electrified Railway Line

Measuring the performance of an electrified railway is particularly complex as it involves a service that requires rolling stock, tracks, safety and signaling systems, stops or stations, and a variety of personnel types. Another factor that affects transport, if it is public, is government intervention to subsidize costs [49]. In the present study, the cost of capital associated with fixed electrical systems has been estimated based on the main components that characterize this scenario:

$$C_I^{ELE} = C_{TPS} + C_{CAT} + C_{MSC} \tag{17}$$

where $C_I^{ELE}$ is the capital cost of the electrification of the railway line, $C_{TPS}$ is the cost associated with the traction power substation, $C_{CAT}$ is the cost associated with the installation of the catenary and $C_{MSC}$ represents other costs related to the electrification intervention such as track lowering in correspondence of tunnels already present in the route and raising of the overpasses. The present value of operation and maintenance (O&M) costs associated with the electrified railway line is calculated as follows:

$$C_{O\&M}^{ELE} = \sum_{n=1}^{N} \frac{C_n^{ELE}}{(1+r)^n} \tag{18}$$

where $C_n^{ELE}$ is the annual operations cost of year $n$ including both fixed and variable costs:

$$C_n^{ELE} = C_e N_d E_d^{TPS} + C_n^{TPS} + C_n^{CAT} \tag{19}$$

where $E_d^{TPS}$ is the energy supplied by the traction power substation in a day, during $N_d$ days within a year, $C_e$ the average cost of electricity, $C_n^{TPS}$ and $C_n^{CAT}$ represent the average of the annual costs associated with the maintenance of the traction power substation and the catenary, respectively, estimated from 1% to 3% per year of the investment cost, referring to the date of commissioning of the equipment [50]. Finally, the annual cost of energy is calculated by using the following equation:

$$ACOE_{ELE} = CRF \cdot \left( C_I^{ELE} + C_{O\&M}^{ELE} \right) \tag{20}$$

### 2.4.2. Costs of Associated with the Use of Trains Equipped with On-Board ESS

The costs taken into account are the capital cost of the on-board ESS, the annual operation costs of the on-board ESS, operation and maintenance (O&M) costs of the on-board ESS and the replacement cost of the on board ESS.

The use of trains equipped with ESS might reduce capital costs as there is no catenary, but the construction of new specialized trains could significantly increase the total costs of realization. Moreover, depending on how the sizing of the energy storage system is carried out, it may be necessary to install a charging system with an installed power comparable to that of a traction power substation, thus affecting capital costs. Furthermore, it is necessary to respect the limits of the state of charge within which the ESS must deliver and absorb energy, even with high discharge current peaks provided they occur for short periods of time, resulting in a useful life of 10÷15 years [51]. However, is needed an expertise in train operation and real work cycle of storage system to know if this useful life is respected.

Consequently, the cost assessment must include the future replacement of the energy storage system following life expectancy and battery life reported in data sheet (not considering battery disposal).

In cost assessment is not taken into account the cost due to the buy of new trains, but is assumed that the cost of the on board ESS is the difference cost between traditional electric trains and new trains equipped with on board ESS. So, the following analysis is relevant on the hypothesis of complete renewal of the rolling stock, as diesel trains.

The cost of capital of the equipment of the storage systems $C_I^{ESS}$, has been estimated as follows:

$$C_I^{ESS} = C_P P_{ESS} + C_E E_{ESS} + C_{FC} \tag{21}$$

where $C_P$ [€/kW] and $C_E$ [€/kWh] are the ESS specific costs, $P_{ESS}$ and $E_{ESS}$ are the power and energy capacities and $C_{FC}$ are the ESS fixed costs associated, for example, with the installation of a recharging system. The present value of operation and maintenance costs associated with the use of trains equipped with on-board ESS $C_{O\&M}^{ESS}$, is calculated by:

$$C_{O\&M}^{ESS} = \sum_{n=1}^{N} \frac{C_n^{ESS}}{(1+r)^n} \tag{22}$$

where $C_n^{ESS}$ is the annual operations cost of year $n$ including both fixed and variable costs, and is computed by using the following equation:

$$C_n^{ESS} = C_f P_{ESS} + C_v N_d E_d^{ESS} + N_d \frac{E_d^{ESS}}{\eta_{ch}} C_{ch} \tag{23}$$

where $C_f$ [€/kW-yr] represents the specific operating costs, $C_v$ [€/kWh] the variable operating costs, $C_{ch}$ [€/kWh] the cost of recharging ESS, $N_d$ the number of days the ESS is active within one year, $P_{ESS}$ [kW] and $E_d^{ESS}$ [kWh] the nominal power and energy supplied by the ESS, and finally $\eta_{ch}$ is the charge efficiency. The replacement cost of the ESS is expressed in the following equation:

$$C_R^{ESS} = C_{FR}\left[ (1+r)^{-LR} + (1+r)^{-2LR} + \dots \right] \tag{24}$$

where $C_{FR}$ represents the future value of replacement cost and $L_R$ is the ESS lifetime. Finally, the annual cost of energy is calculated by using the following equation:

$$ACOE_{ESS} = CRF \cdot \left( C_I^{ESS} + C_{O\&M}^{ESS} + C_R^{ESS} \right) \tag{25}$$

## 3. On-Board ESS Design

In the railway sector, the applications of electrical energy storage systems are usually characterized by the power and energy they must provide. A good sizing of the ESS must allow compliance with the required energy and power constraints, without too many margins.

### 3.1. ESS Specification

The sizing is performed considering a nominal voltage for the battery pack and for the supercapacitor pack, and a power profile assigned for each [28,32]. The subdivision of the original power profile is achieved through a low pass filter and an amplitude limiter [47,52,53]. Several optimization studies have been carried out on this topic [32,35,54]. In this paper, it is considered that the ESS may be composed of high-power lithium-ion cells or high-energy lithium-ion cells and supercapacitors (H-ESS). Shape and weight of the energy storage system are binding constraints since they limit the energy that can be stored on-board. Therefore, the sizing depends on several factors: autonomy and maximum power required; where the charging is performed; the type of storage system.

### 3.2. ESS Sizing

Given a commercial battery cell characterized by a nominal cell voltage $U_{CELL_B}$ the number of cells to be connected in the series $N_{SE_B}$ to obtain a given rated voltage of the battery pack, $U_B$ is obtained by:

$$N_{SE_B} = \frac{U_B}{U_{CELL_B}} \tag{26}$$

It is highlighted that it represents a simplified model for the preliminary study developed in this paper, but in the case of a cell pack, balancing and operating safety should be added. The number of branches to be connected in parallel $N_{PAR_B}$ is determined as the maximum number between $N_{PAR_B}^P$, which represents the number of branches in parallel to satisfy the power requirement to be delivered in traction $P_B^{max}$, and $N_{PAR_B}^E$ which is the one necessary to satisfy the energy required by the battery, $E_B$. In order to increase battery lifetime, the variations in the state of charge are limited between a $SOC_{min}$ of $0.2 \div 0.3$ and a $SOC_{max}$ of $0.8 \div 0.95$. This involves an oversizing with respect to the energy required by the reference cycle but allows to exploit the supply or absorption of high currents for short intervals of time, provided that the fluctuations in the SOC are less than 5% (micro-cycles). An energy storage system of this type, subject to this type of stress, has an expected useful life of $10 \div 15$ years [51,52,55]. Moreover, several studies have been carried out evaluating the use of degraded batteries, since they still have remaining capacity for grid support applications or emergency power supply [56–58]:

$$N_{PAR_B} = \max(N_{PAR_B}^P; N_{PAR_B}^E) \tag{27}$$

$$N_{PAR_B}^E = \frac{E_B}{N_{SE_B} \cdot U_{CELL_B} \cdot C_{AH_{CELL_B}} \cdot (SOC_{max} - SOC_{min})} \tag{28}$$

$$E_B = \int_{t_0}^{t_0+T} P_B(t)\,dt \tag{29}$$

$$N_{PAR_B}^P = \frac{P_B^{max}}{N_{SE_B} \cdot P_{CELL_B}^{max}} \tag{30}$$

where $P^{max}_{CELL_B}$ is the maximum power that the single electrochemical cell can deliver, determined in the discharge phase by using Equation (31), where $R_{DIS}$ [h$^{-1}$] is the discharge C-rate, which is a measure of the rate at which the cell is discharged relative to its nominal capacity $C_{AH_{CELL_B}}$ [Ah]. The maximum power that the single cell can absorb in the charging phase is computed as a function of the charge C-rate $R_{CH}$ [h$^{-1}$] [31]:

$$P_{CELL_{B_{DIS}}} = R_{DIS}{\cdot}C_{AH_{CELL_B}}{\cdot}U_{CELL_B} \tag{31}$$

$$P_{CELL_{B_{CH}}} = R_{CH}{\cdot}C_{AH_{CELL_B}}{\cdot}U_{CELL_B} \tag{32}$$

Given a commercial supercapacitor of rated voltage $U_{CELL_{SC}}$, the number of cells to be connected in series $N_{SE_{SC}}$ to obtain a given rated voltage of the supercapacitor pack $U_{SC}$ is determined by:

$$N_{SE_{SC}} = \frac{U_{SC}}{U_{CELL_{SC}}} \tag{33}$$

The number of supercapacitor branches to be connected in parallel, $N_{PAR_{SC}}$, is determined through Equation (34), where $E_{SC\_max}$ is the maximum energy that the supercapacitor pack must deliver. It is highlighted that the voltage range is limited between $V_{NOM}/2$ (SOC$_{SC}$ = 0) and $V_{NOM}$ (SOC$_{SC}$ = 1), delivering 75% of the stored energy:

$$N_{PAR_{SC}} = \frac{8}{3} \frac{E_{SC\_max}}{\left(N_{SE_{SC}}{\cdot}U_{CELL_{SC}}\right)^2} {\cdot} \frac{N_{SE_{SC}}}{C_{CELL}} \tag{34}$$

$$E_{SC}\left(t_p\right) = \int_{t_0}^{t_0+t_p} [P_{SC}(t) - P_{SC}(t_0)]dt \tag{35}$$

Finally, the total mass $W_{TOT}$ and volume $V_{TOT}$ linked to each energy storage system can be computed as follows:

$$W_{TOT} = (1+\gamma){\cdot}N_{SE}{\cdot}N_{PAR}{\cdot}W_{CELL} \tag{36}$$

$$V_{TOT} = (1+\delta){\cdot}N_{SE}{\cdot}N_{PAR}{\cdot}V_{CELL} \tag{37}$$

where the coefficients $\gamma$ and $\delta$ represent mass and volume rations of DC/DC converters and all the other additional elements necessary for the assembly and use of the energy storage system.

## 4. Simulation Procedures

The railway system model was implemented in a rail simulator based on the 'quasi static' backwards looking method, due to its short simulation times for estimating energy consumption of vehicles following an imposed speed cycle [46]. The rail simulator is a multi-stage program, implemented in Visual Basic and Fortran language and operating in MS-DOS and the Microsoft Excel workspace [59–61], including:

- an electro-mechanical simulator for the evaluation of the rolling stock power consumption for defined traffic scenarios;
- an electrical and thermal simulator to evaluate the energy state of the traction system.

The software allows one to perform high-quality studies from an energy point of view, with a high degree of flexibility in the simulation by means of a graphical interface where it is possible to set many parameters such as trains' departure times, specific train sequences and trains' stop time period in each station.

For the sizing and verification of the energy storage systems, models have been implemented in the Matlab/Simulink environment. Given the parameters of the electrochemical cells and supercapacitors, and given the power profile to be supplied, it is possible to evaluate electrical variables such as voltage, current, state of charge, power and energy supplied and absorbed.

The numerical simulations have been divided into three parts. The first part is related to the train performance simulations, to determine the power profile required by the rolling stock. The second part is related to the traction system simulations, in which the system is studied from an energy point of view, inserting traffic data and imposing simulation constraints. The last part is related to the on-board energy storage systems, which presents the ESS performance given a power profile reference.

*4.1. Train Performance Simulation*

The train performance, computed in the Microsoft Excel environment, requires the main characteristics of the route as input data, namely the plano-altimetric characteristics of the line (slopes), the curvature radius of the curves, speed limits and tunnels. Moreover, it requires electromechanical information of the rolling stock such as: weight, aerodynamic resistance, traction and braking force [46,59]. Given these inputs, and with a calculation step in meters, it is possible to obtain several outputs such as:

- kinematic parameters: travel and braking times, speed and acceleration profiles in time and space;
- dynamic parameters: aerodynamic resistance, resistance associated with curves, slopes and inertia, traction effort at the wheels and power required at the pantograph.

*4.2. Electrical Model of the Traction System*

The electrical computation software, implemented in Fortran language, allows to solve load flows calculation for a direct current electrified system [46,60,61]. For each simulation step, the software creates an equivalent electric network in which the nodes represent the traction power substation, vehicles present on the route and parallel points, as shown in Figure 5. In this model, $V_{TPS}$ and $R_{TPS}$ represent the substation DC voltage and internal resistance; $R_{BIN}$ is the rail electric resistance and $P^{BIN}$ is the electric power required by each train.

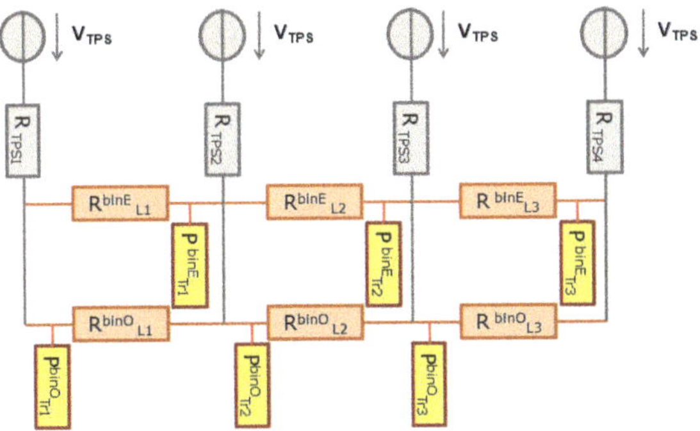


**Figure 5.** Electric software circuit model.

The software requires as input the data related to the power profiles of trains, the electrical substation features and the resistance of contact line equivalent section and rails. As output, the software provides the line voltage profile and currents flowing in the substation feeders, as well as the power provided by each electrical substation [61,62].

*4.3. Electrical Model of the On-Board Energy Storage System*

Simulink/Matlab models have been developed for the battery pack and supercapacitor pack, starting respectively from the characteristics of the electrochemical and supercapacitor cells. The models have been implemented as dynamic continuous-time systems, thus using mainly integrator, sum and gain blocks. In the case of electrochemical cells, a lookup table dynamic block is used to represent the open circuit voltage as a function of the state of charge (OCV-SOC curves). Figure 6 shows the battery based ESS Simulink model.

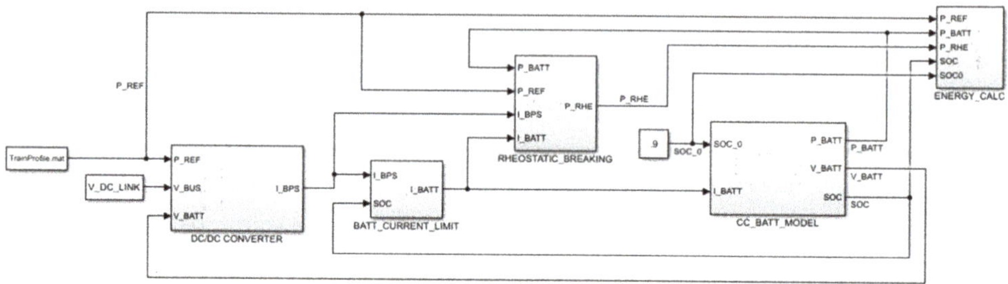

**Figure 6.** Battery based ESS Simulink model.

In addition to the parameters related to the energy storage system, the software requires as input data the reference power profile required, which is obtained directly from the evaluation of the train performance in the case of high-power lithium cells. In the case of H-ESS, the reference power profiles for the high-energy lithium cells and supercapacitors are obtained from a Simulink model that applies a low pass filter and an amplitude limiter to the original power profile.

## 5. Numerical Simulations

### 5.1. Case Study

A real single-track railway line has been chosen as case study. It presents an electrified section at 3 kV DC and a non-electrified section, currently covered by diesel-powered trains. Therefore, passengers are obliged to transfer to a diesel-powered train at the end of the electrified section, in order to reach one of the following stops or stations. The attention is focused to the non-electrified section in order to determine the technical-economic convenience of electrification at 3 kV DC or the use of trains equipped with on-board energy storage systems. Figure 7 reports an overview of the case study.

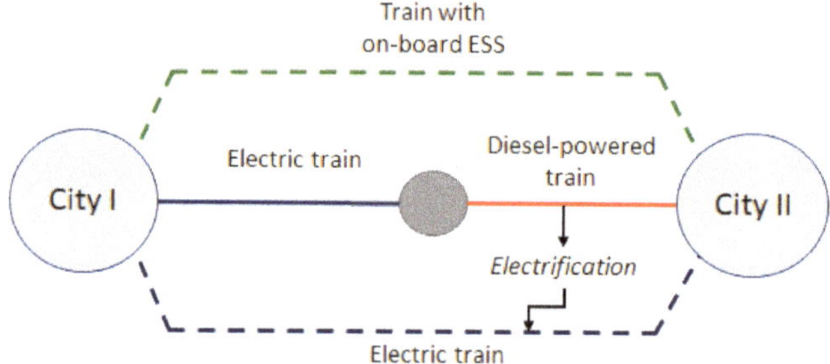


**Figure 7.** Railway connection between City I and City II and possible scenarios.

The non-electrified section is about 16.5 km long and links Station A, at the end of the electrified section with Station B, at the end of the non-electrified section (City II), in about 30 min for each direction. Between Station A and B there are 5 stops, as shown in Figure 8.

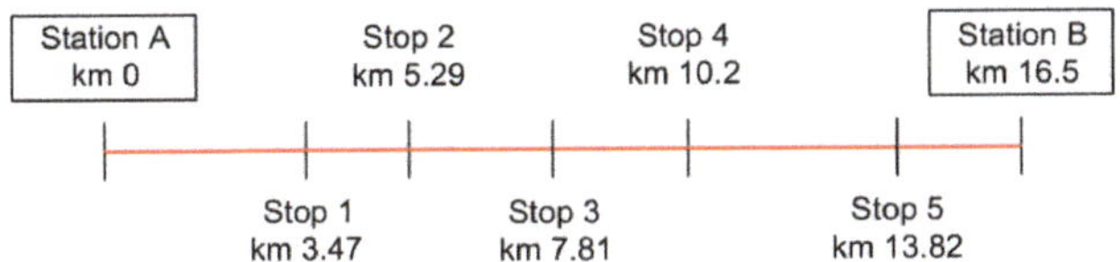



**Figure 8.** Stops between Station A and Station B.

Figure 9 shows the plano-altimetric profile of the railway line. It is characterized by an average slope of 9‰, with a maximum slope of 24‰. There are 22 curves with a minimum bending radius of 200 m.

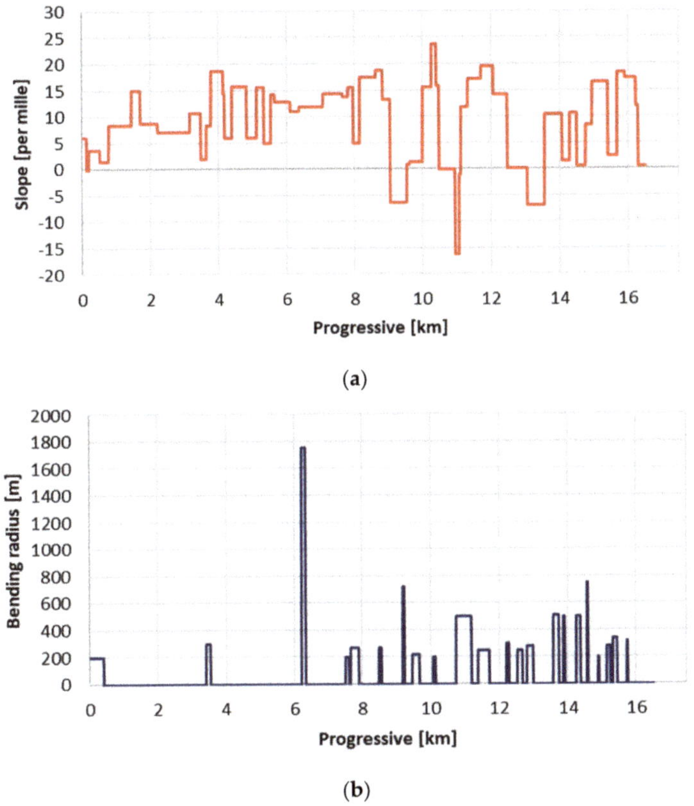



**Figure 9.** Track features: slopes (**a**) and curves (**b**).

For the non-electrified section, the following scenarios have been analyzed: 3 kV DC electrification, since it is the feeding system currently used in the preceding section; high autonomy ESS and ESS with recharge station. For the ESS it has been considered the use of power-oriented Li-ion battery cells (lithium iron phosphate, LFP) and, in the case of hybrid ESS, energy-oriented Li-ion battery cells (lithium nickel manganese cobalt oxide, NMC - lithium nickel cobalt aluminium oxide, NCA) accompanied by supercapacitors.

The electrification at 3 kV DC involves the installation of a catenary and traction power substation, to be located at Stop 2 (progressive km 5.29). Contrariwise, the use of trains equipped with high autonomy on-board ESS allows to minimize the capital costs since it involves the recharge of the battery and supercapacitor pack in the electrified section of the line. However, in this case the ESS weight is high. The installation of a charging infrastructure at Station B (progressive km 16.5), allows reducing the on-board ESS rated capacity and therefore, the total weight of the system. In this paper, it is considered for all the scenarios, the use of the same electric train, whose traction curve is presented in Figure 10.

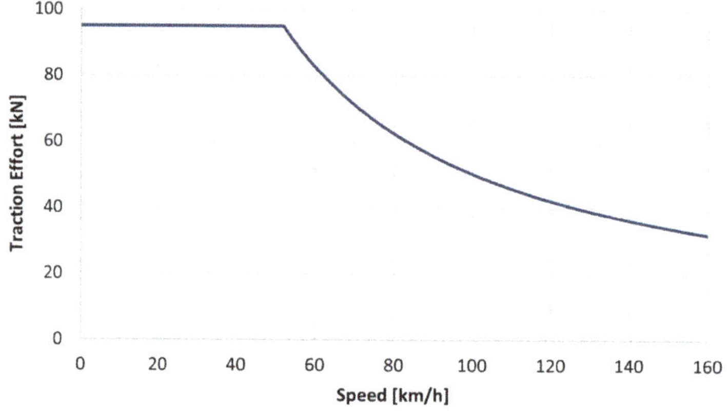


**Figure 10.** Rolling stock traction curve.

The main characteristics of the lithium-ion and supercapacitor cells are illustrated in Tables 1 and 2, respectively. This data is used to size the ESS as reported in the Section 3 of this paper. Table 3 shows the main characteristics of the different components of the traction system: rolling stock, track, DC feeder system and on-board ESS.

**Table 1.** Li-ion cell features.

| Parameter | LithiumWerks ANR26650M1B | Saft MP 176,065 xlr |
|---|---|---|
| Mass [g] | 76 | 150 |
| Rated capacity [Ah] | 2.5 | 6.8 |
| Nominal voltage [V] | 3.3 | 3.65 |
| Specific energy [Wh/kg] | 109 | 165 |
| Max. discharge current [A] | 50 (20C) | 14 (2C) |
| Max. charge current [A] | 10 (4C) | 6.8 (1C) |
| Technology | Power-oriented | Energy-oriented |

**Table 2.** Ultracapacitor cell features.

| Parameter | Maxwell K2 UC–2.85V/3400F |
|---|---|
| Mass [g] | 520 |
| Rated capacity [F] | 3400 |
| Nominal voltage [V] | 2.85 |
| Max. voltage [V] | 3 |
| Specific energy [Wh/kg] | 7.6 |
| Specific power [W/kg] | 8500 |

**Table 3.** Railway System Parameters.

| Parameter | Value |
| --- | --- |
| **Rolling Stock** | |
| Net weight | 184 t |
| Loaded weight | 256 t |
| Rotating mass coefficient | 1.05- |
| Max. train speed | 160 km/h |
| Nominal deceleration | 0.8 m/s$^2$ |
| Accessories power | 200 kW |
| Power train overall efficiency | 0.8- |
| **Track** | |
| Track's length | 16.5 km |
| Von Röckl formula coef. A | 0.65 m |
| Von Röckl formula coef. B | 55 m |
| **DC Feeder System** | |
| Catenary section | 320 mm$^2$ |
| Rail electric resistance | 0.083 Ω/km |
| Substation location | 5.29 km |
| Substation nominal power | 3.6 MVA |
| AC/DC conversion units | 2- |
| Substation DC voltage | 3000 V |
| Substation internal resistance | 0.2 Ω |
| Maximum line voltage (+20%-CEI EN 50163) | 3600 V |
| Minimum line voltage (−33%-CEI EN 50163) | 2000 V |
| **On-board ESS** | |
| DC link nominal voltage | 3000 V |
| Max. (Min.) battery SOC value | 90 (30)% |
| Battery pack nominal voltage | 1500 V |
| Supercapacitor pack nominal voltage | 1000 V |
| DC/DC efficiency | 95% |

## 5.2. Results and Discussion

Several simulations are carried out in order to determine the technical and economic convenience of the proposed scenarios, using the procedures reported on Section 4.

### 5.2.1. Train Performance

The train performance has been performed for each direction of travel: from Station A to Station B and vice versa. Figure 11 shows the speed profile from Station A to Station B, highlighting that the train that travels the line from one stop to another, accelerating until the speed limit is reached and decelerating accordingly of stops.

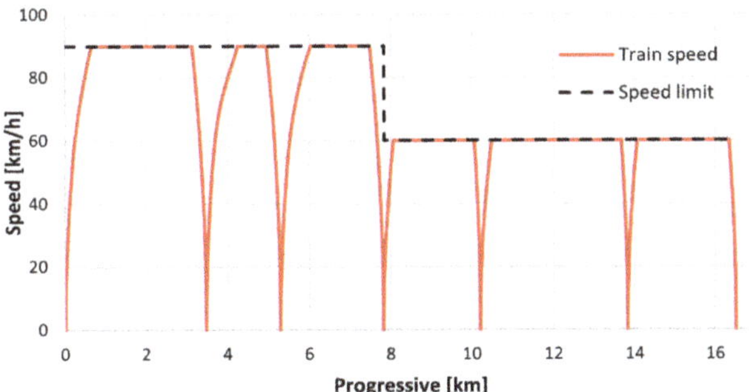


**Figure 11.** Speed profile from Station A to Station B.

Starting from the speed profile, it is possible to obtain the power profile of each direction of travel, as shown in Figures 12 and 13. For each single section present between one stop and another, the train uses all the effort theoretically available only in the acceleration phases until the limit speed of the track is reached. Subsequently, in the steady state phase, the tractive effort applied at the wheels follows the trend of the resistance offered by the various accidentalities present in the track, influencing the power requested by the train. It is also noted that the maximum power at the pantograph in the traction phase is 3.7 MW and the maximum power in the regenerative braking phase is 2 MW. During stops, the only power consumed by the train is related to auxiliary services.

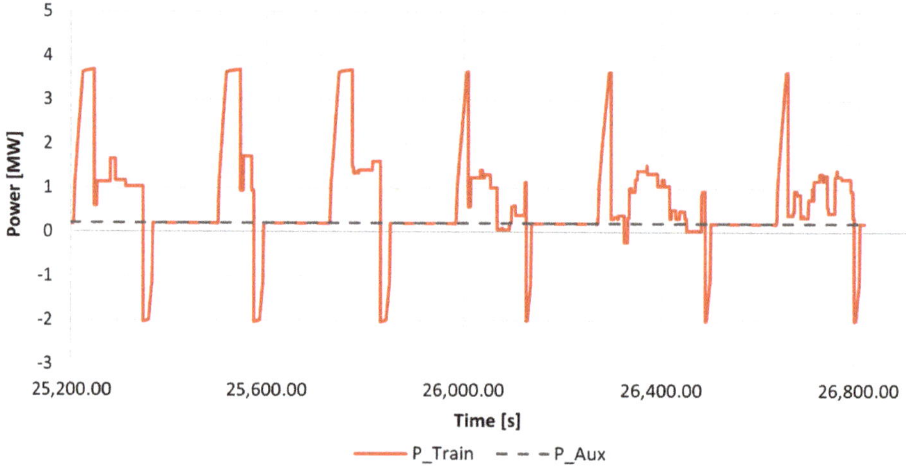

**Figure 12.** Power profile from Station A to Station B.

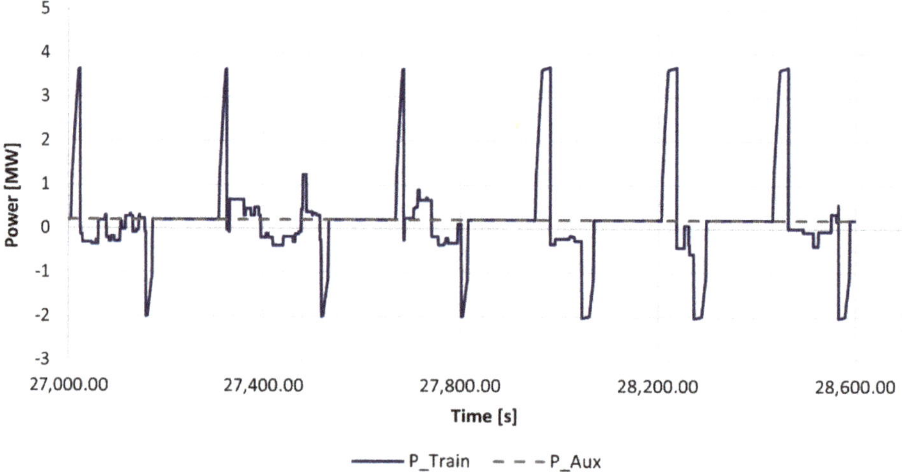

**Figure 13.** Power profile from Station B to Station A.

From Station A to Station B, the energy consumption is about double respect of the opposite direction, where the regenerative braking energy is high as the rolling stock must brake to contain its speed. It is due to the negative sign of the track resistance since the train goes downhill.

Figure 14a shows the energy consumption of the train and the regenerative braking energy, for each direction of travel. Figure 14b reports the timetable of the railway line under study (26 min and 51 s from Station A to Station B and 26 min, and 38 s in the opposite direction).

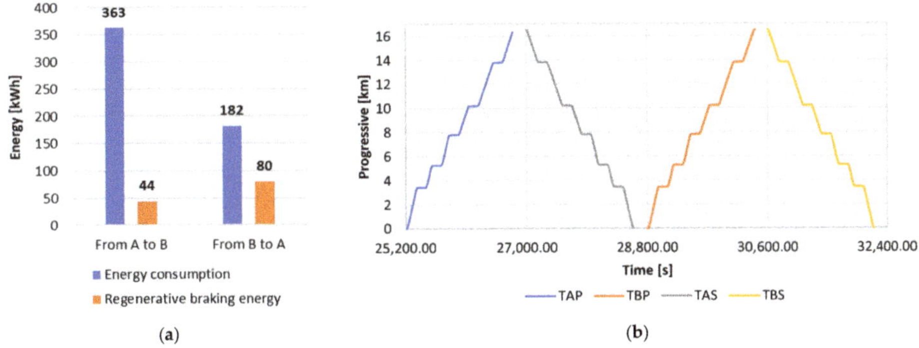

**Figure 14.** Energy required by trains: (**a**) Energy data; (**b**) Traffic conditions of the railway line.

## 5.2.2. Electrification at 3 kV DC

The electrification of the section under study include the installation of a traction power substation and the contact line. Therefore, it is necessary to determine the nominal power of the transformer groups present in the substation and the catenary to be used, considering the limits allowed by the CEI EN 50163 standards. For the TPS, located at Stop 2, it has been analyzed the use of two different types of substations: 2 AC/DC conversion units with a nominal power of 3.6 MVA or 5.4 MVA each. For the catenary, it has been considered the use of an equivalent section of 320 mm$^2$ or 440 mm$^2$. For simplicity purposes, only the numerical results related to the 2 × 3.6 MVA TPS with a 320 mm$^2$ are here reported.

Figure 15 reports the minimum line voltage, highlighting that the voltage drop along the line is critical only at 7:30 and 8:30, at the departure of the train from Station B, towards Station A. In this case, the TPS must feed the train that is more than 11 km away. In fact, it is necessary to limit the power to the pantograph to 3 MW in order to contain the voltage drop on the line, obtaining a voltage at the pantograph is 2.19 kV. However, this minimum voltage is present only for a single sample of the simulation (10 s).

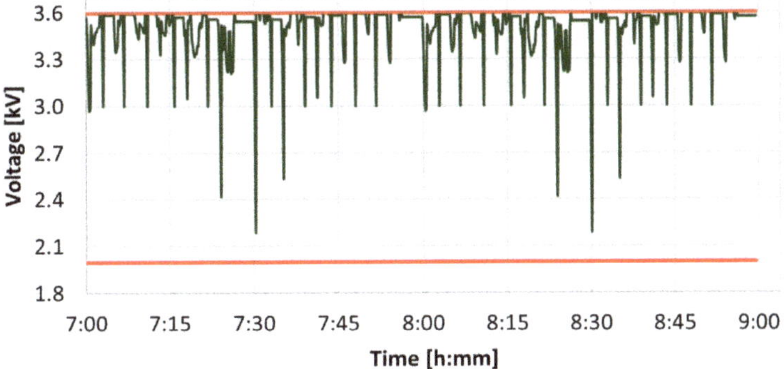

**Figure 15.** Minimum line voltage.

Figure 16 shows the TPS power absorption and given that the power required to the primary network does not exceed 10 MVA, it is possible to provide for the medium voltage connection to the primary network.

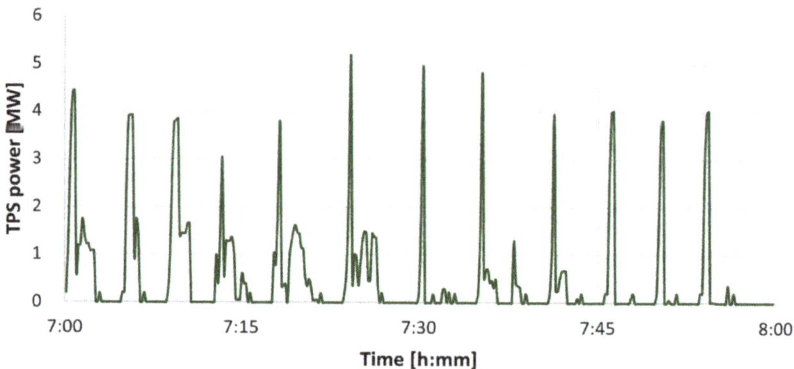

**Figure 16.** Power absorbed by the traction power substation between 7 a.m. and 8 a.m.

### 5.2.3. On-Board ESS

The main constraints of the on-board ESS sizing are the energy required by the cycle and the maximum power to be delivered. In the case of an electrochemical storage system, this must provide both the energy required by the cycle and the power during peaks related to accelerations. Instead, in the case of a H-ESS consisting of batteries and supercapacitors, the power profile required by the train is divided into two parts, obtaining a reference profile for the battery pack and another for the supercapacitor pack.

The high autonomy ESS implies an autonomy to go from Station A to Station B and recharge the storage system at Station A. According to this cycle, the energy to be supplied to the rolling stock is equal to 585 kWh. First, the use of high-power lithium-ion cells has been analyzed, whose characteristics are shown in Table 1. In this case, the sizing results in a 980 kWh (650 Ah) battery pack, of which only 586 kWh can be used, due to SOC limits. In Figure 17, $P_{BATT}$ represents the power associated with the battery pack, positive in the traction phase, and negative in the braking phase. The voltage of the battery pack is identified as $V_{BATT}$ and presents variations that re less than ±5%. The supplied current is identified as $I_{BATT}$ and is positive in the discharge phase (traction) and negative in the charging phase (regenerative braking). Since the capacity of the battery pack (C) is equal to 650 Ah, it is observed that both the maximum discharge and charge current are much lower than the maximum allowed by the battery pack (20C and 4C respectively, in the discharge phase and charge). Moreover, it is observed that the sizing is appropriate because the SOC does not drop too low, and over time (despite greater discharges), the SOC will not drop below a certain minimum (30%).

Afterwards, a high autonomy H-ESS has been analyzed. It is composed of the high-energy lithium-ion and supercapacitor cells, whose characteristics are shown in Tables 1 and 2. The subdivision of the original power profile is achieved through a low pass filter with a cut-off frequency of 1 Hz and an amplitude limiter which limits the maximum power supplied by the batteries to reference value equal to 2 MW, in order to reduce the maximum power peak that they must supply to about 50% of the original one. Instead, the regenerative braking power is limited in absolute value to about 5% of the maximum braking power (0.1 MW), in order to make the supercapacitors recover as much energy as possible during the braking phase. In this way two reference profiles are obtained to carry out the sizing: P_REF_B for the battery pack and P_REF_SC for the supercapacitors, as shown in Figure 18.

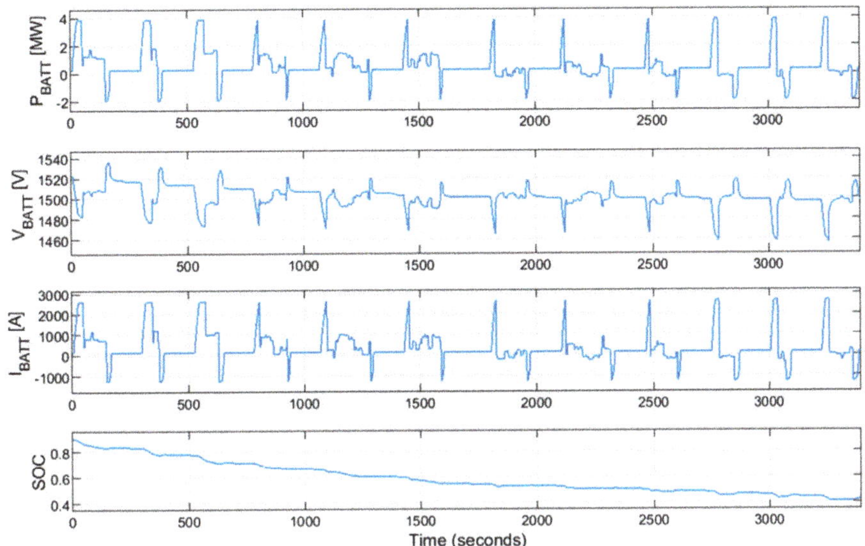

**Figure 17.** High autonomy ESS parameters-High-power lithium ion cells.

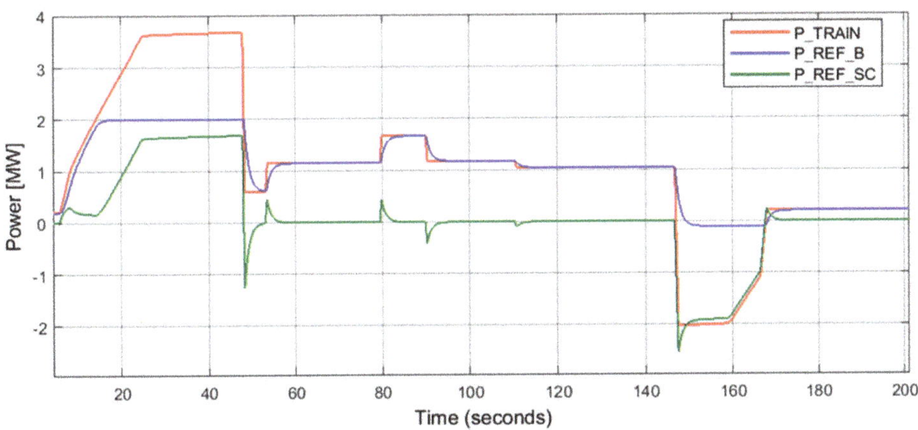

**Figure 18.** Power profiles H-ESS.

Carrying out the verification simulations of the H-ESS, it is observed in Figure 19a that the current required to the battery pack (in blue) tends to exceed the limits set by the manufacturer during discharge and is therefore limited to a lower current (in orange). Moreover, the supercapacitor SOC reaches too low values, as shown in Figure 19b. To avoid these situations, it is necessary to increase the quantity of parallel branches of high-energy lithium and supercapacitors cells, further increasing the total mass of the storage system. This hybrid solution is thus abandoned because it is too heavy.

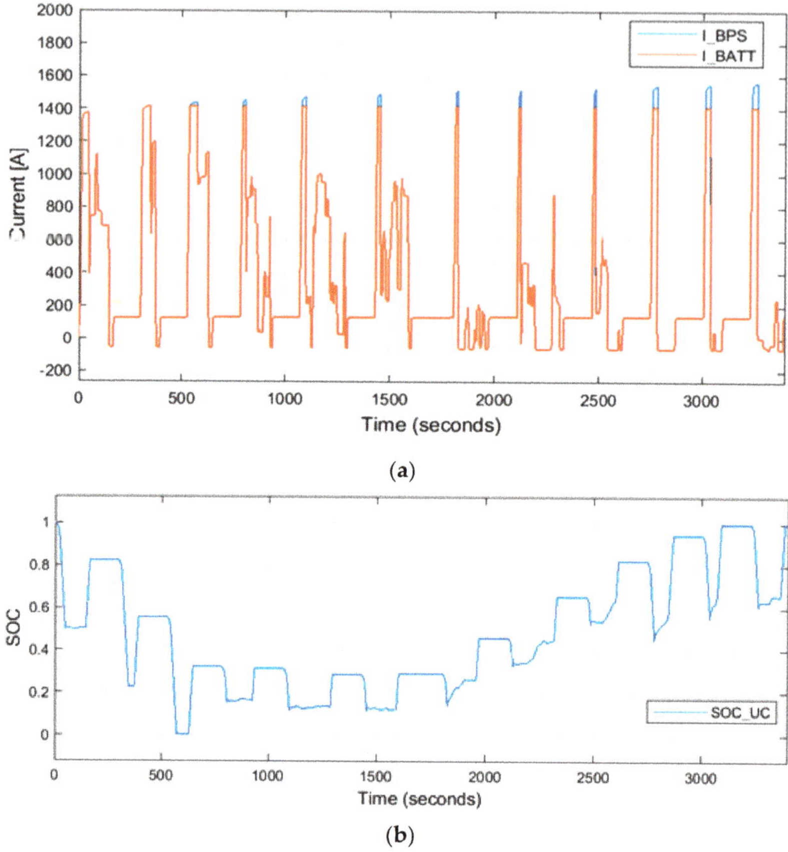

**Figure 19.** H-ESS: (**a**) Current required to the battery pack; (**b**) Supercapacitor pack SOC.

The ESS with recharge station implies an autonomy to go from Station A to Station B, where the recharge needs to be performed. According to this cycle, the energy to be supplied to the rolling stock is equal to 383 kWh. It has been considered the use of high-power lithium-ion cells and the sizing procedure results in a 640 kWh (425 Ah) battery pack, which does not exceed the limits set by the manufacturer. The recharge at Station B can be performed from the 30% to the 50% of the SOC, since it allows arriving to Station A within the available SOC range. Therefore, the recharge must have a duration of 6 min, at 2C rate, implying a recharge station of at least 1.4 MW.

In Figure 20 it is shown the charging time for both high autonomy ESS and ESS with recharge station, to be performed at Station A (in the electrified section). This recharge takes the SOC from 30% to 90% and it is performed, at 2C, in 18 min (LFP cells). Table 4 presents a synthesis of the viable on-board ESS scenarios, including the ESS impact in the energy required by the rolling stock because of the weight increase.

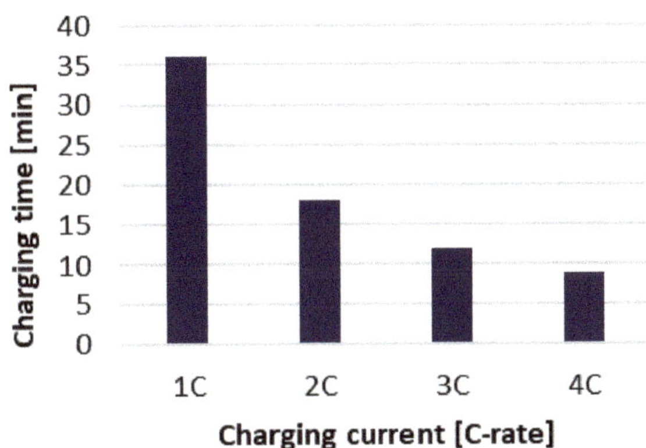

**Figure 20.** On-board ESS charging time.

**Table 4.** On-board ESS scenarios.

|  | High Autonomy ESS | ESS with Recharge Station |
|---|---|---|
| Rated capacity [Ah] | 650 | 427.5 |
| Rated capacity [kWh] | 976 | 642 |
| Available capacity [kWh] | 586 | 385 |
| ESS weight [t] | 9 | 6 |
| Train Performance energy impact [%] | 3 | 2 |
| Recharge power @2C (C-rate) [MW] | 2.2 | 1.4 |

### 5.2.4. Economic Comparison

As reported in Section 2, the present value of total cost and the annual cost of energy are computed, considering the following hypothesis: an expected lifetime of 30 years; ESS lifetime of 10 years; an interest rate of 4% for electrification and 6% for ESS [45]; an average cost of electricity of 15 €/MWh [49]; charge efficiency equal to 90%; four daily connections between Station A–Station B. In Figure 21 it is shown the total cost (a) and the ACOE of the proposed scenarios. 3 kV DC electrification shows a present value of the total costs of approximately €13 M and an ACOE of over 700,000€ per year. Most of these costs come from the high investment cost necessary for the electrification, estimated at €10 M, as it is necessary to install a traction power substation, a contact line and carry out works such as track lowering in correspondence of tunnels already present in the route and raising of the overpasses. The ESS with recharge station has a present value of the total cost of approximately €7 M, considering the replacement of the storage system every 10 years (it will depend on the correct operation in charge/discharge phases), and the installation of the necessary system for charging. The ACOE of this solution is about 500,000€ per year, due to the lower costs associated with the charging station compared to those associated with the installation of TPS. Finally, the high autonomy ESS, with a total cost of €5.5 M and ACOE of 400,000€ per year, could be advantageous solution from the economic point of view, since it does not involve the installation of any infrastructure for the recharge.

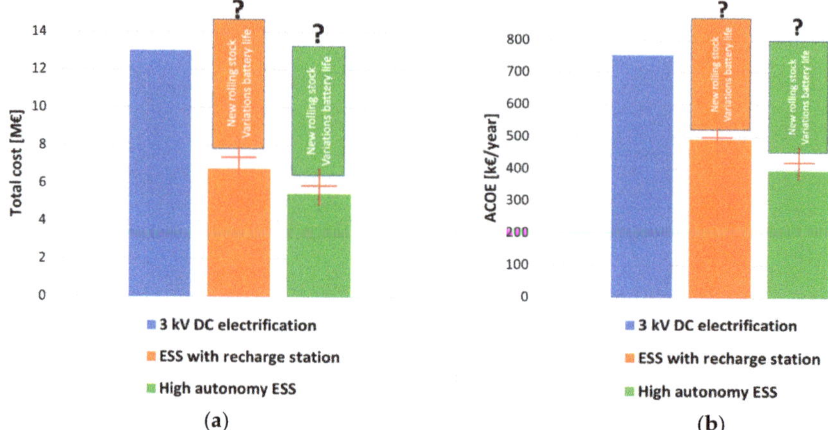


(**a**)  (**b**)

**Figure 21.** Different scenarios comparison: (**a**) Present value of total cost; (**b**) ACOE.

The main advantage of on-board ESS with high autonomy is that it does not require the installation of additional charging systems as this is provided for in the already electrified section of the railway line. Consequently, investment costs are limited. Moreover, on-board storage systems work with shallower charge and discharge cycles.

It should be emphasized, once again, that apparently the total costs linked to the solutions related to the use of trains with on-board storage would seem to be lower but it is necessary to underline that the total cost will strongly depend on the costs related to the construction of new rolling stock. Subsequently the real lifecycle of the batteries could lead to a further increase in costs due to their early replacement. However, investments in e-mobility lead to an increasing mass production of batteries. Consequently, if high demand growth is sustained, battery solutions are expected to be more profitable.

The feasibility of trains with on board ESS is similar to that of fully electrifying the ships that will sail the North Sea in the near future [63]. In both cases, current reliable solutions (i.e., electrified railway lines and heavy fuel oil powered shipping vessels, respectively) may lead to a previous hybridization stage before achieving the complete on-board energy supply system.

## 6. Conclusions

In the very near future more technical-economic comparisons between diesel/electrified railway and the use of trains with on-board energy storage systems will be carried out in order to define the best solution achievable from an environment and economic point of view in relation to different railways lines. It will be necessary define the adoption of these systems for expansion of already electrified railways, point-to-point connections, commuter transport systems and so on.

The times are becoming ripe and consequently the paper defines methodologies, procedures, simulation models considering the main economic aspects, in order to increase the scientific literature on these innovations that will be increasingly investigated.

The methodology and the model developed in this paper could help to define a series of aspect to consider when is necessary compare different system adoptable to modernize a railway line.

The paper shown numerical simulations on a real railway line that presents a 3 kV DC electrified section and a non-electrified section, currently covered by diesel-powered trains. Different types of ESS have been analyzed, evaluating the use of high-power lithium cells or high-specific lithium cells and supercapacitors. The models of the battery pack and supercapacitor pack have been implemented in the Matlab/Simulink environment. Three main scenarios have been evaluated for the non-electrified section: 3 kV DC electrification, on-board ESS with high autonomy and on-board ESS with recharge station.

The results showed that the use of trains equipped with on-board energy storage system could be a good solution, but from a preliminary economic point of view, today many cost parameter are still not well known. Consequently, over the years, through the first realizations and the monitoring of the operation, it will be possible to definite clarification of the costs, as well as they are well established and consolidated in the electrification at 3 kV DC.

It is also shown that the most suitable storage technology for the case under study is that of high-power lithium ion batteries. The on-board ESS with high autonomy presents an ACOE of about €400 k/year, apparently 40% less than the 3 kV DC electrification scenario (costs of new train manufacture, battery disposal are not taken into account). Since the case study concerns a single-track railway line in which maximum speeds are limited, two trains allow to cover the entire section without any impact on the service offered to passengers, given the charging times and autonomy of the chosen solution.

Finally, it should be noted that the electrification of railway lines represents a consolidated and highly reliable solution. Contrariwise, the use of trains equipped with on-board energy storage systems represents a solution currently being tested, which will take many years before a consolidation.

It will be very important to know the cost of the new trains equipped with ESS on board, taking into account the long homologation procedure until they are placed on the market, and the real behavior of the storage systems in relation to life expectancy.

**Author Contributions:** Conceptualization, R.L., A.R., G.G.B. and N.C.; methodology, R.L. and A.R.; software, R.L. and A.R.; validation, R.L. and A.R.; investigation, M.T.; data curation, M.T., A.R., G.G.B. and N.C.; writing—original draft preparation, M.T. and A.R.; writing—review and editing, M.T., A.R., G.G.B. and N.C.; visualization, R.L. and A.R. All authors have read and agreed to the published version of the manuscript.

**Funding:** This research received no external funding.

**Conflicts of Interest:** The authors declare no conflict of interest.

## Abbreviations

| | | | | |
|---|---|---|---|---|
| $ACOE$ | Annual cost of energy | | $E_B$ | Energy of battery |
| $ACOE_{ELE}$ | Annual cost of energy of electrification | | $E_d^{TPS}$ | Energy supplied by the traction power substation in a day |
| $ACOE_{ESS}$ | Annual cost of energy of energy storage system | | $H\text{-}ESS$ | Hybrid energy storage system |
| $C_{AH_{CELL_B}}$ | Nominal capacity of the cell | | $N_{SE_B}$ | Number of cells to be connected in the series |
| $C_I^{ELE}$ | Capital cost of electrification | | $N_{PAR_B}$ | Number of branches to be connected in parallel |
| $C_I^{ESS}$ | Capital cost of on-board ESS | | $N_{SE_{SC}}$ | Number of cells to be connected in series for supercapacitors |
| $C_n^{ELE}$ | Annual operation costs of electrification | | $N_{PAR_{SC}}$ | Number of branches to be connected in parallel for supercapacitors |
| $C_n^{ESS}$ | Annual operation costs of on-board ESS | | $OCV$ | Open circuit voltage |
| $C_{O\&M}^{ELE}$ | Operation and maintenance (O&M) costs of electrification | | $O\&M$ | Operation and maintenance |
| $C_{O\&M}^{ESS}$ | Operation and maintenance (O&M) costs of on-board ESS | | $P_{CELL_B}^{max}$ | Maximum power that the single electrochemical cell can deliver |
| $C_R^{ESS}$ | Replacement cost of the on-board ESS | | $P_{CELL_{B_{CH}}}$ | Maximum power that the single cell can absorb in the charging phase |
| $CRF$ | Capital recovery factor | | $P_{CELL_{B_{DIS}}}$ | Maximum power that the single cell can give in the discharging phase |
| $C_n^{TPS}$ | Annual costs associated with the maintenance of the traction power substation | | $R_{DIS}$ | The discharge C-rate |
| $C_n^{CAT}$ | Annual costs associated with the maintenance of the catenary | | $SC$ | Supercapacitor |
| $C_e$ | The average cost of electricity | | $SOC$ | State of charge |
| $ESS$ | Energy storage system | | $TPS$ | Traction power substation |
| $E_{SC}$ | Energy of supercapacitor | | $U_{CELL_{SC}}$ | Supercapacitor rated voltage |

## References


1. International Energy Agency. Rail. Available online: https://www.iea.org/reports/rail (accessed on 10 September 2020).
2. Electric vehicles from life cycle and circular economy perspectives. In *TERM 2018: Transport and Environment Reporting Mechanism (TERM) Report*; European Environment Agency: Luxembourg, 2018; Available online: https://www.eea.europa.eu/publications/electric-vehicles-from-life-cycle (accessed on 10 September 2020).
3. The First and Last Mile—The Key to Sustainable Urban Transport. In *Transport and Environment Report 2019*; EEA—European Environment Agency: Luxembourg, 2019; Available online: https://www.eea.europa.eu/publications/the-first-and-last-mile. (accessed on 10 September 2020).
4. Messagie, M.; Boureima, F.-S.; Coosemans, T.; Macharis, C.; Mierlo, J.V. A Range-Based Vehicle Life Cycle Assessment Incorporating Variability in the Environmental Assessment of Different Vehicle Technologies and Fuels. *Energies* **2014**, *7*, 1467–1482. [CrossRef]
5. UIC—International Union of Railways, «Railway Statistics Synopsis». Available online: https://uic.org/IMG/pdf/uic-statistics-synopsis-2019.pdf (accessed on 8 September 2020).
6. Kawamoto, R.; Mochizuki, H.; Moriguchi, Y.; Nakano, T.; Motohashi, M.; Sakai, Y.; Inaba, A. Estimation of $CO_2$ Emissions of Internal Combustion Engine Vehicle and Battery Electric Vehicle Using LCA. *Sustainability* **2019**, *11*, 2690. [CrossRef]
7. Hill, N. Impact Analysis of Mass EV Adoption and Low Carbon Adoption Intensity Fuels Scenarios. In Proceedings of the 13th Concawe Symposium, Antwerp, Belgium, 18–19 March 2019.
8. Kamargianni, M.; Li, W.; Matyas, M.; Schafer, A. A Critical Review of New Mobility Services for Urban Transport. *Transp. Res. Procedia* **2016**, *14*, 3294–3303. [CrossRef]
9. Cimen, S.G.; Gnilka, M.; Schmuelling, B. Economical, social and political aspects of e-mobility in Germany. In Proceedings of the 2014 Ninth International Conference on Ecological Vehicles and Renewable Energies (EVER), Monte, Torino, Italy, 15–16 June 2014; pp. 1–5. [CrossRef]
10. Franzò, S.; Chiaroni, D.; Chiesa, V.; Frattini, F. Emerging business models fostering the diffusion of E-mobility: Empirical evidence from Italy. In Proceedings of the 2017 International Conference of Electrical and Electronic Technologies for Automotive, Torino, Italy, 15–16 June 2017; pp. 1–5. [CrossRef]
11. Siemens. A New Era of Sustainable Road Freight Transport. 2020. Available online: https://new.siemens.com/global/en/products/energy/medium-voltage/solutions/emobility/emobility-latest-technologies.html (accessed on 20 September 2020).
12. Felez, J.; García-Sanchez, C.; Lozano, J.A. Control Design for an Articulated Truck with Autonomous Driving in an Electrified Highway. *IEEE Access* **2018**, *6*, 60171–60186. [CrossRef]
13. Xiao, S.; Zhang, C.; Luo, Y.; Wu, J.; Rao, Y.; Sykulski, J.K. A Study on the Detachment Characteristics of the Tramwave Catenary-Free Electrification System for Urban Traffic. *IEEE Trans. Plasma Sci.* **2020**, *48*, 3670–3678. [CrossRef]
14. Bombardier Transportation. Solutions and Technologies. 2020. Available online: https://www.bombardier.com/en/transportation/products-services.html (accessed on 20 September 2020).
15. Mayer, L. *Impianti Ferroviari*; CIFI: Rome, Italy, 2016; Volume 2.
16. Miyatake, M.; Haga, H.; Suzuki, S. Optimal speed control of a train with On-board energy storage for minimum energy consumption in catenary free operation. In Proceedings of the 2009 13th European Conference on Power Electronics and Applications, Barcelona, Spain, 8–10 September 2009; pp. 1–9.
17. Al-Ezee, H.; Sarath, T.; Taylor, I.; Daniel, S.; Schweickart, J. An On-board Energy Storage System for Catenary Free Operation of a Tram. *Renew. Energy Power Qual. J.* **2017**, *1*, 540–544. [CrossRef]
18. Steiner, M.; Klohr, M.; Pagiela, S. Energy storage system with ultracaps on board of railway vehicles. In Proceedings of the2007 European Conference on Power Electronics and Applications, Aalborg, Denmark, 2–5 September 2007; pp. 1–10. [CrossRef]
19. Ghaviha, N.; Campillo, J.; Bohlin, M.; Dahlquist, E. Review of Application of Energy Storage Devices in Railway Transportation. *Energy Procedia* **2017**, *105*, 4561–4568. [CrossRef]
20. Ogura, K. Test Results of a High Capacity Wayside Energy Storage System Using Ni-MH Batteries for DC Electric Railway at New York City Transit. In Proceedings of the 2011 IEEE Green Technologies Conference (IEEE-Green), Baton Rouge, LA, USA, 14–15 April 2011; pp. 1–6. [CrossRef]
21. Available online: https://www.globalmasstransit.net/archive.php?id=15973 (accessed on 8 September 2020).

22. Liu, X.; Li, K. Energy storage devices in electrified railway systems: A review. *Transp. Saf. Environ.* **2020**, *2*, 183–201. [CrossRef]

23. Hayashiya, H.; Masuda, M.; Noda, Y.; Suzuki, K.; Suzuki, T. Reliability analysis of DC traction power supply system for electric railway. In Proceedings of the 2017 19th European Conference on Power Electronics and Applications (EPE'17 ECCE Europe), Warsaw, Poland, 11–14 September 2017; pp. P.1–P.6. [CrossRef]

24. White, R.D. AC 25kV 50 Hz electrification supply design. In Proceedings of the 6th IET Professional Development Course on Railway Electrification Infrastructure and Systems (REIS 2013), London, UK, 3–6 June 2013; pp. 108–150. [CrossRef]

25. Fei, Z.; Konefal, T.; Armstrong, R. AC railway electrification systems—An EMC perspective. *IEEE Electromagn. Compat. Mag.* **2019**, *8*, 62–69. [CrossRef]

26. Jafri, N.H.; Gupta, S. An overview of Fuel Cells application in transportation. In Proceedings of the 2016 IEEE Transportation Electrification Conference and Expo, Asia-Pacific (ITEC Asia-Pacific), Busan, Korea, 1–4 June 2016; pp. 129–133. [CrossRef]

27. Rizzo, R.; Spina, I.; Boscaino, V.; Miceli, R.; Capponi, G. A Novel Fuel Cell-Based Power System Modeling Approach. In Proceedings of the 2013 European Modelling Symposium, Manchester, UK, 20–22 November 2013; pp. 370–374. [CrossRef]

28. Graber, G.; Galdi, V.; Calderaro, V.; Piccolo, A. Sizing and energy management of on-board hybrid energy storage systems in urban rail transit. In Proceedings of the 2016 International Conference on Electrical Systems for Aircraft, Railway, Ship Propulsion and Road Vehicles & International Transportation Electrification Conference (ESARS-ITEC), Toulouse, France, 2–4 November 2016; pp. 1–6. [CrossRef]

29. Iwase, T.; Kawamura, J.; Tokai, K.; Kageyama, M. Development of battery system for railway vehicle. In Proceedings of the 2015 International Conference on Electrical Systems for Aircraft, Railway, Ship Propulsion and Road Vehicles (ESARS), Aachen, Germany, 3–5 March 2015; pp. 1–6. [CrossRef]

30. Unique Energy Hub. Universal Ragone Plot. Available online: https://uenergyhub.com/info/universal-ragone-plot/ (accessed on 6 September 2020).

31. Sadoun, R.; Rizoug, N.; Bartholomeüs, P.; Barbedette, B.; Le Moigne, P. Optimal sizing of hybrid supply for electric vehicle using Li-ion battery and supercapacitor. In Proceedings of the 2011 IEEE Vehicle Power and Propulsion Conference, Chicago, IL, USA, 6–9 September 2011; pp. 1–8. [CrossRef]

32. Mesbahi, T.; Rizoug, N.; Bartholomeüs, P.; Sadoun, R.; Khenfri, F.; Le Moigne, P. Optimal Energy Management for a Li-Ion Battery/Supercapacitor Hybrid Energy Storage System Based on a Particle Swarm Optimization Incorporating Nelder–Mead Simplex Approach. *IEEE Trans. Intell. Veh.* **2017**, *2*, 99–110. [CrossRef]

33. Ostadi, A.; Kazerani, M.; Chen, S. Optimal sizing of the Energy Storage System (ESS) in a Battery-Electric Vehicle. In Proceedings of the 2013 IEEE Transportation Electrification Conference and Expo (ITEC), Detroit, MI, USA, 16–19 June 2013; pp. 1–6. [CrossRef]

34. Sadoun, R.; Rizoug, N.; Bartholomeüs, P.; Le Moigne, P. Optimal architecture of the hybrid source (battery/supercapacitor) supplying an electric vehicle according to the required autonomy. In Proceedings of the 2013 15th European Conference on Power Electronics and Applications (EPE), Lille, France, 2–6 September 2013; pp. 1–7. [CrossRef]

35. Cao, J.; Emadi, A. A new battery/ultra-capacitor hybrid energy storage system for electric, hybrid and plug-in hybrid electric vehicles. In Proceedings of the 2009 IEEE Vehicle Power and Propulsion Conference, Dearborn, MI, USA, 7–11 September 2009; pp. 941–946. [CrossRef]

36. Ostadi, A.; Kazerani, M. A Comparative Analysis of Optimal Sizing of Battery-Only, Ultracapacitor-Only, and Battery–Ultracapacitor Hybrid Energy Storage Systems for a City Bus. *IEEE Trans. Veh. Technol.* **2015**, *64*, 4449–4460. [CrossRef]

37. Tummakuri, V.; Chelliah, T.R.; Ramesh, U.S. Sizing of Energy Storage System for A Battery Operated Short Endurance Marine Vessel. In Proceedings of the 2020 IEEE International Conference on Power Electronics, Smart Grid and Renewable Energy (PESGRE2020), Cochin, India, 2–4 January 2020; pp. 1–6. [CrossRef]

38. Balsamo, F.; Lauria, D.; Capasso, C.; Veneri, O.; Pede, G. Main issues with the design of batteries to power full electric water busses. In Proceedings of the 2016 AEIT International Annual Conference (AEIT), Capri, Italy, 5–7 October 2016; pp. 1–6. [CrossRef]

39. Bolonne, S.R.A.; Chandima, D.P. Sizing an Energy System for Hybrid Li-Ion Battery-Supercapacitor RTG Cranes Based on State Machine Energy Controller. *IEEE Access* **2019**, *7*, 71209–71220. [CrossRef]

40. Brenna, M.; Foiadelli, F.; Stocco, J. Battery Based Last–Mile Module for Freight Electric Locomotives. In Proceedings of the 2019 IEEE Vehicle Power and Propulsion Conference (VPPC), Hanoi, Vietnam, 14–17 October 2019; pp. 1–6. [CrossRef]

41. Weiss, H.; Winkler, T.; Ziegerhofer, H. Large lithium-ion battery-powered electric vehicles—From idea to reality. In Proceedings of the 2018 ELEKTRO, Mikulov, Czech Republic, 21–23 May 2018; pp. 1–5. [CrossRef]

42. Royston, S.J.; Gladwin, D.T.; Stone, D.A.; Ollerenshaw, R.; Clark, P. Development and Validation of a Battery Model for Battery Electric Multiple Unit Trains. In Proceedings of the IECON 2019—45th Annual Conference of the IEEE Industrial Electronics Society, Lisbon, Portugal, 14–17 October 2019; pp. 4563–4568. [CrossRef]

43. Construcciones y Auxiliar de Ferrocarriles (CAF) S.A. Greentech, CAF's Solution to the Catenary-Free Tram. Available online: https://www.caf.net/en/ecocaf/nuevas-soluciones/tranvia-greentech.php (accessed on 6 September 2020).

44. Bombardier Transportation. World Premiere: Bombardier Transportation Presents a New Battery-Operated Train and Sets Standards for Sustainable Mobility. 2018. Available online: https://rail.bombardier.com/en/newsroom/press-releases.html/bombardier/news/2018/bt-20180912-world-premiere-bombardier-transportation-presents-a/en (accessed on 6 September 2020).

45. Bombardier Transportation. E-Mobility and Battery Technology. 2020. Available online: https://rail.bombardier.com/en/solutions-and-technologies/urban/e-mobility-battery-technology.html (accessed on 6 September 2020).

46. Graber, G.; Calderaro, V.; Galdi, V.; Piccolo, A.; Lamedica, R.; Ruvio, A. Techno-economic Sizing of Auxiliary-Battery-Based Substations in DC Railway Systems. *IEEE Trans. Transp. Electrif.* **2018**, *4*, 616–625. [CrossRef]

47. Lamedica, R.; Ruvio, A.; Galdi, V.; Graber, G.; Sforza, P.; GuidiBuffarin, G.; Spalvieri, C. Application of battery auxiliary substations in 3kV railway systems. In Proceedings of the 2015 AEIT International Annual Conference (AEIT), Naples, Italy, 14–16 October 2015; pp. 1–6. [CrossRef]

48. Ruiz-Cortés, M.; Romero-Cadaval, E.; Roncero-Clemente, C.; Barrero-González, F.; González-Romera, E. Energy management strategy to coordinate batteries and ultracapacitors of a hybrid energy storage system in a residential prosumer installation. In Proceedings of the 2017 International Young Engineers Forum (YEF-ECE), Almada, Portugal, 5 May 2017; pp. 30–35. [CrossRef]

49. Lee, J.; Ahn, J.; Lee, B.K. A novel li-ion battery pack modeling considering single cell information and capacity variation. In Proceedings of the 2017 IEEE Energy Conversion Congress and Exposition (ECCE), Cincinnati, OH, USA, 1–5 October 2017; pp. 5242–5247. [CrossRef]

50. Gattuso, D.; Restuccia, A. A Tool for Railway Transport Cost Evaluation. *Procedia Soc. Behav. Sci.* **2014**, *111*, 549–558. [CrossRef]

51. Baumgartner, J.P. Prices and Costs in the Railway Sector. Laboratoire d'Intermodalité des Transports Et de Planification 2001, Lausanne. Available online: http://archiveweb.epfl.ch/litep.epfl.ch/files/content/sites/litep/files/shared/Liens/Downloads/Divers/Baumgartner_Couts_chf_2001_e.pdf (accessed on 6 September 2020).

52. Ceraolo, M.; Giglioli, R.; Lutzemberger, G. Prove Sperimentali e Ciclo Semplificato di un Sistema di Accumulo Per Una Tramvia. ENEA—National Agency for New Technologies. 2015. Available online: https://www.enea.it/it/Ricerca_sviluppo/documenti/ricerca-di-sistema-elettrico/accumulo/2014/rds-par2014-175.pdf (accessed on 6 September 2020).

53. Marignetti, F. *La Trazione Ferroviaria: I Sistemi a Guida Vincolata*; Società Editrice Esculapio: Bologna, Italy, 2018.

54. Ehsani, M.; Yimin, G.; Emadi, A. *Modern Electric, Hybrid Electric and Fuel Cell Vehicles Fundamentals, Theory and Design Second Edition*; CRC Press: Boca Raton, FL, USA, 2010.

55. Chu, H.; Wang, X.; Mu, X.; Gao, M.; Liu, H. Research on Optimal Configuration of Hybrid Energy Storage Capacity Based on Improved Particle Swarm Optimization Algorithm. In Proceedings of the 2019 IEEE 4th Advanced Information Technology, Electronic and Automation Control Conference (IAEAC), Chengdu, China, 20–22 December 2019; pp. 2030–2036. [CrossRef]

56. LeBel, F.; Trovao, J.P.; Boulon, L.; Sari, A.; Pelissier, S.; Venet, P. Lithium–Ion Cell Empirical Efficiency Maps. In Proceedings of the 2018 IEEE Vehicle Power and Propulsion Conference (VPPC), Chicago, IL, USA, 27–30 August 2018; pp. 1–4. [CrossRef]

57. Calderaro, V.; Galdi, V.; Graber, G.; Lamberti, F.; Piccolo, A. A sizing method for economic assessment of II-life batteries for power system applications. In Proceedings of the 2017 IEEE Power & Energy Society General Meeting, Chicago, IL, USA, 16–20 July2017; pp. 1–5. [CrossRef]

58.  D'Orazio, A.; Elia, S.; Santini, E.; Tobia, M. Succor System and Failure Indication for the Starter Batteries of Emergency Gensets. *Period. Polytech. Electr. Eng. Comput. Sci.* **2020**, *64*, 412–421. [CrossRef]

59.  Boccaletti, C.; Elia, S.; Salas, M.E.F.; Pasquali, M. High reliability storage systems for genset cranking. *J. Energy Storage* **2020**, *29*, 101336. [CrossRef]

60.  Capasso, A.; Lamedica, R.; Ruvio, A.; Giannini, G. Eco-friendly urban transport systems. Comparison between energy demands of the trolleybus and tram systems. *Ing. Ferrov.* **2014**, *69*, 329–347.

61.  Capasso, A.; Ceraolo, M.; Lamedica, R.; Lutzemberger, G.; Ruvio, A. Modelling and simulation of tramway transportation systems. *J. Adv. Transp.* **2019**, *2019*, 4076865. [CrossRef]

62.  Lamedica, R.; Gatta, F.M.; Geri, A.; Sangiovanni, S.; Ruvio, A. A software validation for dc electrified transportation system: A tram line of Rome. In Proceedings of the 2018 IEEE International Conference on Environment and Electrical Engineering and 2018 IEEE Industrial and Commercial Power Systems Europe (EEEIC/I&CPS Europe), Palermo, Italy, 12–15 June 2018; pp. 1–6. [CrossRef]

63.  Savard, C.; Nikulina, A.; Mécemmène, C.; Mokhova, E. The Electrification of Ships Using the Northern Sea Route: An Approach. *J. Open Innov. Technol. Mark. Complex.* **2020**, *6*, 13. [CrossRef]

**Publisher's Note:** MDPI stays neutral with regard to jurisdictional claims in published maps and institutional affiliations.

MDPI

St. Alban-Anlage 66

4052 Basel

Switzerland

Tel. +41 61 683 77 34

Fax +41 61 302 89 18

www.mdpi.com

*Energies* Editorial Office

E-mail: energies@mdpi.com

www.mdpi.com/journal/energies